Fish and Fisheries of Himalayan and Terai Region of West Bengal with Ornamental Touch

by

Dr. Madhumita Mukherj
Director of Fisheries
Govt. of West Bengal

R.F. Lepcha
Joint Director of Fisheries (HQ), West Bengal

Dr. Chinmoy Chakraborty
Assistant Director of Fisheries, West Bengal

2011

DAYA PUBLISHING HOUSE®

Delhi - 110 035

ISBN 81-7035-699-7
ISBN 978-81-7035-699-8

Cover Design by: Mehatab Molla

Published by : **Daya Publishing House®**
1123/74, Deva Ram Park
Tri Nagar, Delhi - 110 035
Phone: (011) 27383999
Fax: (011) 23260116
e-mail : dayabooks@vsnl.com
website : www.dayabooks.com

Showroom : 4760-61/23, Ansari Road, Darya Ganj,
New Delhi - 110 002
Phone: 23245578, 23244987

Laser Typesetting : **Classic Computer Services**
Delhi - 110 035

Printed at : **Chawla Offset Printers**
Delhi - 110 052

PRINTED IN INDIA

Acknowledgement

Authors take this precious moment to express their thanks to Sri Kiranmoy Nanda, Hon'ble MIC, Department of Fisheries, Aquaculture, Aquatic Resources and Fishing Harbours, Government of West Bengal, for his adnascent encouragement and prudential approach in planning, execution and presentation of this book and eventually steering it to a successful endeavor. Authors express their deep sense of gratitude towards Sri S. Ghising, Hon'ble Chairman, DGHC and the Sri D. T. Tamlong, IAS, Principal Secretary, DGHC for providing the necessary facilities to carry out the research work.

Authors can not but bewray their deep sense of indebtedness towards Sri R.P.S. Kahlon, IAS, Secretary, Department of Fisheries, Aquaculture, Aquatic Resources and Fishing Harbours, Government of West Bengal for his all-round encouragement and necessary direction during execution and presentation of this work. Authors are obliged to Sri S. Bhattacharya, IAS, Director of Fisheries, Government of West Bengal for his help and co-operation.

Authors acknowledge the sincere co-operation received from Sri S. Chakrabarti, Joint Director of Fisheries (ME & MS), Dr. P. K. Jana, Deputy Director of Fisheries (ME & MS) and Dr. U. Sar, Deputy Director of Fisheries (Northern Zone) for providing relevant fisheries data of North Bengal.

Authors have pleasure to express their gratitude towards Govt. of India for providing necessary fund for the research work in the field of Coldwater Fisheries of the state.

Heartfelt thanks to all the departmental officers of districts head quarters and blocks of Darjeeling, Jalpaiguri, Coochbehar, Uttar Dinajpur, Dakshin Dinajpur and Malda for their keen co-operation during planning of this book. Authors also owe a word of thanks to all the Research Fellows under Coldwater Fisheries Project, local people and fishers of North Bengal for their immense cooperation and help during the study. We also acknowledge Pritha Sarkar and Mujibar Rahaman, Sayantani Roy, Research Fellow, ARHMC for data computation of this book.

Needless to mention, errors and omissions are responsibility of authors.

Dr. Madhumita Mukherjee

R.F. Lepcha

Dr. Chinmoy Chakraborty

Preface

The scenic beauty of North Bengal has always attracted many tourists, academicians and even researchers year after year. It has vast resources in terms of forests, mountains and terrain regions. Because of its unique diversity and topographical features of the landscape, the North Bengal is blessed with numerous flora and fauna. Beside, tea gardens, forests, hills, it is also bestowed with numerous big and small rivers and tributaries *viz.* Teesta, Rangeet, Balasun, Mahananda etc. which are flowing through North Bengal.

For the last decade, the Department of Fisheries, Aquaculture, Aquatic Resources and Fishing Harbours, GOWB, with the collaboration of DGHC is coming up with numbers of fishery projects for sustainable utilisation of its vast resources and playing a vital role towards the upliftment and shaping of the socio-economic condition of the people of this region. However, the innumerable aquatic resources of Northern part of the state is being utilized for the development of freshwater fishery resources to a large extent, but a meager attention has been paid for the development of aquaculture and fisheries specially in coldwater region. No authenticated data is available regarding the catch structure of various fisheries from the area and very little work is done on scientific lines to study the topography, temperature regime and wholesome ichthyo-fauna of the rivers. There is no such coldwater fish farm or any co-ordinate research centre to have first hand information of this region. No authentic data base survey has been taken up for quantification of state coldwater resources and *vis–a–vis* assessment of their production potential for over all future planning and development of this virgin area.

The important fish species of this region are on the verge of decline due to various interferences of man and natural calamities. A detailed survey on the water quality, fish species diversity of the North Bengal with special emphasis to coldwater fisheries is therefore important for propagation and conservation of the important flora and fauna of this region in order to come up with proper planning for sustainable exploitation without disturbing the natural eco-system. The Department of Fisheries, GOWB with the funding support from GOI has implemented a Project to explore the resource and breeding ground of coldwater species.

Hence an attempt has been made in this book to assess the resource inventory of coldwater fisheries, to study the growth, food, feeding habit and viability of coldwater fish species in confined water, to assess the constraints of implementing coldwater fishery and to survey the present fisheries scenario of North Bengal and measures to be taken in a systematic manner by the Department of Fisheries, Aquaculture, Aquatic Resources and Fishing Harbours, GOWB.

Dr. Madhumita Mukherjee

R.F. Lepcha

Dr. Chinmoy Chakraborty

Contents

List of Figures

List of Plates

List of Tables

Chapter 1
Introduction

The northern part of the state of West Bengal, particularly the Himalayan range has been bestowed with a great wealth of perennial rivers and small streams. These rivers and tributaries, namely, the Teesta, Rangeet, Ralli, Rishi, Reyang, Gelkhola Pala, Mahananda, Balasun and others are sources of important sport and ornamental fish species, crabs and prawns which have high export value. These precious species have never been investigated properly nor have they been identified for commercial purposes. So far no research has been carried out on biological productivity of these upland rivers holding cold-water biotic community. There is no substantial data of cold-water fisheries to come up with credible information to provide a firm base for formulating suitable guidelines for conservation, propagation and judicious exploitation of the existing resources of the cold-water fisheries in the state of West Bengal. While fishery sector in the plains has achieved considerable progress in fish production techniques – both in capture and culture fisheries; the development of fishery in the hill regions have remained by a large stagnant. This can be attributed to the poor socio-economic condition and lack of proper fishing techniques among the hill people as well as lack of research and proper scientific data.

The number of indigenous, flora and fauna is declining every year due to the various man made and natural calamities. Moreover, a number of mini hydroelectricity dams are under construction across the river Teesta that will further add to the decline of various fish species. The rampant destruction of different species of fish during spawning time especially in small streams and tributaries by the local people with the use of pesticides and chemicals and the destruction of their breeding ground with the help of dynamite has taken its toll. Viable strategies and options for increasing and sustaining coldwater fishery sector have to be developed to safe guard and conserve these cold-water fisheries of this State and to ensure food security in the near future. The administrative units, population structure and rainfall of different districts of North Bengal are revealed in Tables 1 –3.

Table 1: Administrative Unit in North Bengal

Sl.No.	Name of the District	Sub-division	Police Station	Develop-ment Block	Pan-chayet Samiti	Gram Pan-chayet	No. of Village	Munici-palities	Muni-cipal Corp.
1.	Darjeeling	4	16	12	4	134	708	4	1
2.	Jalpaiguri	18	70	69	61	721	7606	20	1
3.	Coochbihar	5	10	12	12	128	1202	6	0
4.	Uttar Dinajpur	2	9	9	9	99	1504	5	0
5.	Dakshin Dinajpur	2	8	8	8	65	1638		0
6.	Malda	2	11	15	15	147	1798	2	0
7.	Total	33	124	125	109	1294	14456	37	2
8.	West Bengal	65	446	341	333	3360	40782	117	6
9.	Per cent Share of North Bengal	50.77	27.80	36.66	32.73	38.51	35.45	31.62	33.33

Source: Census 2001.

Table 2: Population by Sex in North Bengal

Sl.No.	Name of the District	Area in Sq Km	Male	Female	Total	Rural	Urban
1.	Darjeeling	3149	826334	779566	1605900	1085023	520877
2.	Jalpaiguri	6227	1753278	1649926	3403204	2799357	603847
3.	Coochbihar	3387	1271715	1206565	2478280	2252774	225506
4.	Uttar Dinajpur	3140	1260747	1181077	2441824	2147353	294471
5.	Dakshin Dinajpur	2219	770443	732204	1502647	1306004	196643
6.	Malda	3733	1689409	1600751	3290160	3049245	240915
7.	Total	21855	7571926	7150089	14722015	12639756	2082259
8.	West Bengal	88752	41487694	38733477	80221171	57734690	22486481
9.	Per cent Share of North Bengal	24.62	18.25	18.46	18.35	21.89	9.26

Source: Census 2001.

Table 3: Annual Rainfalls (mm) in North Bengal

Sl.No.	Name of the District	Normal	1990	1997	1998	1999	2000	2001
1.	Darjeeling	3212	3810	2966	3474	3323	3123	2727
2.	Jalpaiguri	4136	3434	2935	4698	3736	3733	3205
3.	Coochbihar	3193	3157	2530	4141	3182	2961	2526
4.	Uttar Dinajpur	1802	2416	2027	3026	2811	1955	2259
5.	Dakshin Dinajpur		2402	1464	2267	2306	1455	1656
6.	Malda	1498	2159	1719	2059	2141	1500	1557

Source: Directorate of Agriculture, GOWB; and Meteorological Dept., GOI.

Our country, India has very rich aquatic biodiversity spread across different ecosystems. Out of 27,800 fish species reported globally, about 11 per cent (2182) species are found in Indian waters. To achieve sustainable utilization of these resources, appropriate planning for conservation and management is needed. The hilly areas in India are abode of great fish diversity and about 3.3 per cent of our fish resources live in upland waters. Coldwater fisheries occupy an important place amongst the freshwater fishes of India. However the ever increasing demand for fish and their indiscriminate capture in the past have caused considerable damage to the natural upland fishery resources. Most Indian upland rivers are almost heavily fished. Added to these, the high degree of alternative use of water, mainly for power generation and agriculture influence the structure of the eco-system.

It is reported that at least 131 species of fishes are found in the Darjeeling hills and the adjoining Terai and Duars region. Of late, it would be very difficult to collect even 50-60 species from this area, which points to a very concerning situation possibly due to various reasons:

(*i*) Soil erosion and landslides are the direct fallout of indiscriminate deforestation and lead to increasing sediment load of the riverbeds, which in its turn create natural barrier for the migration of fishes.

(*ii*) Besides massive quantities of sediments eroded from cleared watersheds are eventually discharged into the rivers and steams, where they can be extremely destructive to the aquatic organisms.

(*iii*) Indiscriminate use of different pesticides and other chemicals in the Tea plantations ultimately find their way into the aquatic systems.

(*iv*) Construction of the Teesta Barrage over the Teesta near Millanpalli village (Jalpaiguri district) has severely hampered the migration of the fishes from the plains to the upland and vice versa. Many such hydral projects are under implementation which will further result in declining of the indigenous species. This may result in permanent and irrevocable reduction in fish stocks.

A detailed survey on the water quality, fish species diversity of the rivers of North Bengal and its tributaries is therefore important for propagation and conservation of the important flora and fauna of this region in order to come up with proper planning for sustainable exploitation and without disturbing the natural eco-system (Chakraborty, 2007b). The Department of Fisheries, GOWB with the funding support from GOI had implemented a project to explore the resource and breeding ground of coldwater species. The main objectives of this study were as follows:

(*i*) Assessment of the resource inventory of coldwater fisheries.

(*ii*) Culture of coldwater fish species in confined water and study the growth, survival rate, food and feeding habit and its viability.

(*iii*) Assess the constraints of implementing coldwater fishery in the running water fishpond.

(*iv*) Survey the present status of the fish farmers of the hills, to come up with suitable recommendations.

Chapter 2

Aquatic Resources and Potentials of North Bengal

Culture of fish in the North Bengal districts has largely concentrated on the production of stocking material for rivers and streams, and some lakes. In the districts of North Bengal the cultivation of fish contributes little to the overall freshwater fish production. Virtually every facility created for fish cultivation in the districts of North Bengal for stocking material to meet up the shortage of seed requirement. The commercially important sport fishery is also dependent to the larger extent on the stocking with artificially breed fry and fingerlings.

Coldwater aquaculture in the Indian Himalayas has been closely associated with the introduction of exotic trouts and common carps. The lack of fast-growing indigenous fish in the Himalayas motivated the British administrators in India to transplant exotic species, such as trout and other salmonids, from Europe to meet their need for recreational fishing.

Today, the available technologies allow the culture of a number of exotic and indigenous coldwater fish species in the Indian Himalayas. The most common exotic species are rainbow trout, brown trout, common carp, grass carp while the indigenous fish are: mahseers (*Tor putitora* and *Tor tor*), and schizothoracines (*S. progastus, Schizothorax richardsonii, S. niger* and *S. curvifrons*). Among these *Tor putitora, N. hexagonolepis, Labeo dero, Labeo gonesis, S. progastus* and *S. richardsonii* are preferred because of their wide range of distribution in the Himalayas.

Golden mahseer (*Tor putitora*) is a widely distributed fish, present from Hindukush-Kabul-Kohistan in the northwest to Sadiya in the eastern Himalayas, as well as in Myanmar and Thailand. The species, a multiple spawner, migrates annually upstream to a higher altitude to feed and spawn. The snow trout or mountain barbell

(*Schizothorax richardsonii*) is widely distributed in the Indian Himalayas, from Ladakh in the northwest to Sadiya in the east. The species is an inhabitant of snow-melt and glacier-fed streams in the Greater and Lesser Himalayas. They undertake migration during winter months when the temperature in the Greater Himalayan waters reaches the near-freezing point.

2.1. Fishery Status of North Bengal

Most of the major wetlands are on the verge of extinction. Increasing population seems to be at the root of problems causing environmental changes – most like causes being regular floods, soil erosion, etc. An increase in human settlements in these low lying areas – [subject to flood and reclamation of the wetlands on either side of the rivers as it facilitates cultivation] has been reducing their flood absorption capacity. District-wise agriculture features of North Bengal have been shown in Plate 1. Many people are losing their prime agricultural land, mango orchard, including homelands as well due to heavy flood situation. To minimize the effect of flood, conservation plan of these major wetlands have to be taken up with the following action plan:

(*i*) In cooperation with the local govt. office, large wetlands can be converted into piscicultural system for the cultivation of fish. Fish seeds, hatchlings are available locally in plenty which can be utilized.

(*ii*) Farmers losing their prime agricultural land can produce fruits like Makhana, water chestnut etc.

(*iii*) Extra water for irrigation could be obtained from wetlands during dry season.

2.2. Aquatic Resources

The districts of North Bengal (Figure 1) of this state is bestowed with innumerable aquatic resources which are to a large extent being utilized for the development of freshwater fishery resources, but a meager attention has been paid for the development of coldwater fisheries. Three hill sub-divisions namely Darjeeling, Kalimpong and Kurseong of Darjeeling district has main potentials in coldwater fisheries. Interspread with innumerable small and large lotic systems the district has a culturable water area of about 1537 ha and 3398 km of riverine resources (Ghoshal 2002). Blessed with numerous rivers and springs *i.e.* Teesta, Relli, Balasun, Mahanada, Jaldhaka, Tarkhola, Rishi, Pala Riyang holds fish wealth.

Amongst 131 fish species identified from Darjeeling and Jalpaiguri districts, only 21 are presently reported (Barat *et al.*, 2005). According to Mukherjee *et al.* (2005), about 125 fish species have been recorded from Darjeeling hill area, some of which are found in torrential streams and are remarkably well adopted for clinging to rocks in swift currents. From zoo-geographical view point, the fish fauna of this region is of special interest because it is the converging of Chinese, Malayan and Indian sub-continent of Oriental region. The major fish fauna of economic importance reported from these uplands includes Katli (*Neolissocheilus hexagonolepis*), Mahseer (*Tor putitora* and *T. tor*), Goonch (*Bagarius bagarius*), Gurdi (*Labeo dero* and *L. dyocheilus*), Indian trout (*Raimas bola*), Asla (*Schizothorax richardsonii*) whereas the fish fauna of less

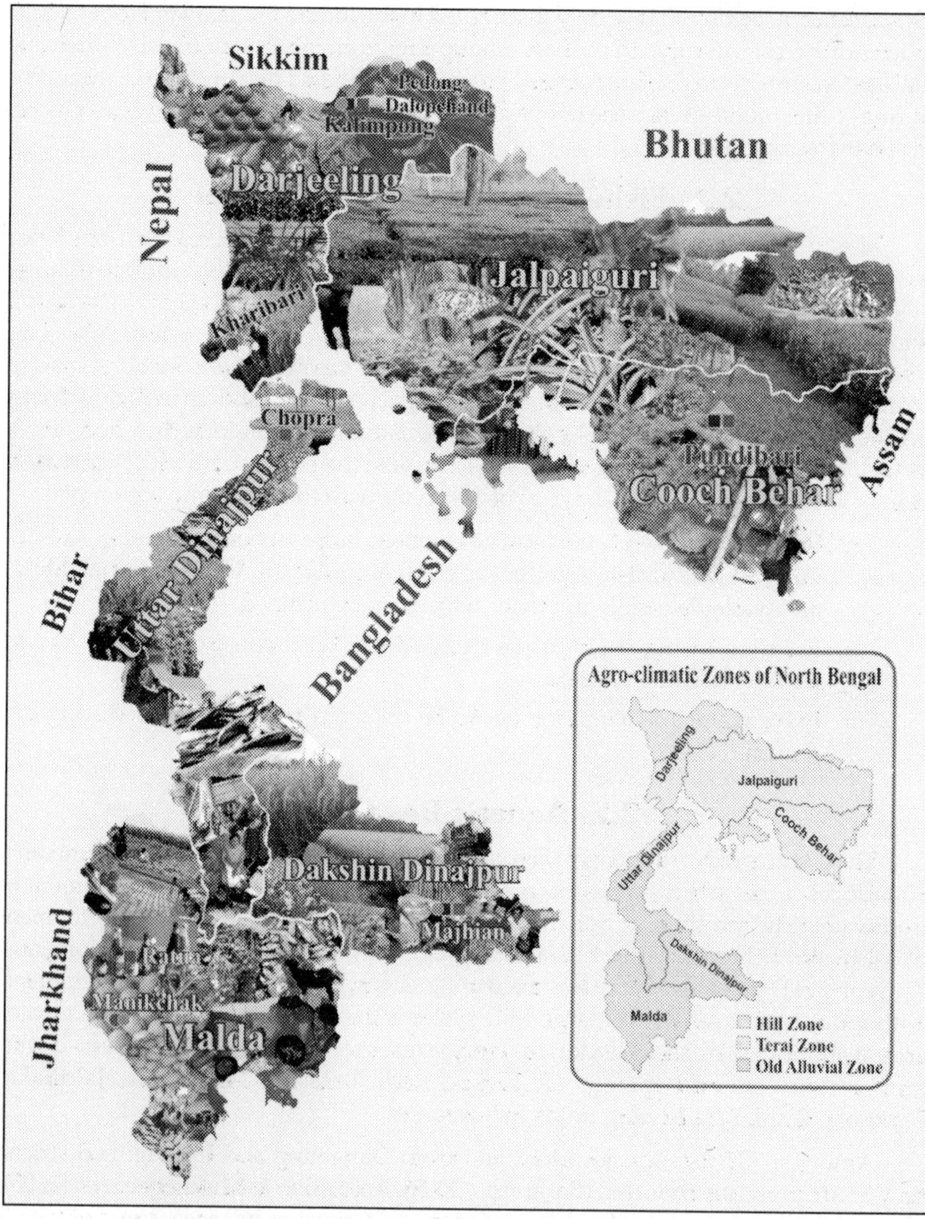

Plate 1: District-wise Agriculture Features of North Bengal

commercial importance comprise Budena (*Garra gotyla*), Utta (*Labeo pangusia*), Faketa (*Barilius bendelisis*), Taktaka (*Barilius barna*), Lohari (*Crossocheilus latius*), Chepti (*Cyprinion semiplotus*), Bhitti (*Danio aquipinnatus*), Nipati (*Danio dangila*), Tilkaprey (*Euchiloglanis hodgarti*) and Titakaberi (*Balitoria brucei*). According to Jhingran and Sehgal (1978), attempts to establish brown trout in Darjeeling streams from Kashmir could not be successful.

Source: www.mapsofindia.com

Figure 1: District Map of West Bengal

Probably no authenticated data is available regarding the catch structure of various fisheries from the area as Ghoshal (*op. cit.*) states that little work is done on scientific lines to study the topography, temperature regime and wholesome fish fauna of the rivers. So no coldwater fish farm or any co-ordinate research centre has any first hand information of the region. No authentic data base survey has been taken up for quantification of state coldwater resources and for rearing of sub-temperate fishes and their production potential. This is required to be undertaken for over all future planning and development of this virgin area.

2.3. Declining Trends of Aquatic Species

There have been reports that important coldwater fish fauna in Darjeeling and surrounding regions are showing depletion in natural stocks and some of candidates are even critically endangered due to multitude of varied anthropogenic interferences, of which the major ones include: (*a*) ecological erosions, (*b*) construction of Teesta barrage, (*c*) immense use of insecticides and pesticides, (*d*) unchecked poaching, (*e*) bunding of streams and (*f*) Heavy landslides in 1968 resulting rise of river beds, disappearance of good number of permanent pools harbouring game fishes.

Besides all these factors, even after degeneration of existing stocks of coldwater fisheries, the nature exhibits diversity and holds enough prospects for fisheries development under planned programme and conservation. All sections of hill streams especially river Teesta harbours considerable fishery of mahseers and goonch, which are generally fished by rod and lines especially at the confluence of smaller stream with the main river. The chocolate mahseer is still widely distributed in Darjeeling hills. The dominance of *Neolissocheilus hexagonolepis* over other coldwater fish species could be attributed to i) very long spawning period (May-June, August–September), ii) unique and swift migratory habit iii) ability to withstand stress and iv) undergo spawning in different environmental conditions.

2.4. Jhora Fishery

Darjeeling, the paradise of scenic beauty and centre of global nostalgia on fascinating aroma of tea, is called as "Queen of the hills" which lies in the Northern part of the State of West Bengal in India, roughly resembles an inverted wedge with its resting on Sikkim, its sides touching Nepal, Bhutan and the Jalpaiguri district of West Bengal. Geographically it lies between latitude 26° 27″S and 27° 13″N and longitude 88° 53″ E and 89° 59″E and area 3,149 sq.kms. The Darjeeling district is divided in to four Sub-divisions namely Darjeeling (Sadar), Kalingpong, Kurseong and Siliguri comprising of 12 (Twelve) blocks *viz.* (*i*) Darjeeling-Phulbazar (23); (*ii*) Rangli-Rangliot (11), (*iii*) Jor.-Sukhiapokhari (16), under Darjeeling Sadar; (*iv*) Kalimpong-I (18) (*v*) Kalimpong-II (13), (*vi*) Gorubathan (11), under Kalimpong; (*vii*) Kurseong (14), (*viii*) Mirik (6), under Kurseong; (*ix*) Matigara (5), (*x*) Naxalbari (6), (*xi*) Kharibari (4), and (*xii*) Phansidewa (7) under Siliguri where the number of Grampanchyat have been defined in parenthesis. Besides, the district is declared as industrially backward by the State Government and is eligible for various concessions. As per 2001 Census, a large chunk of population in district constitutes non-workers (65.78 per cent). In these hill areas there are plenty numbers of hill streams. This encouraging result prompted the State Fisheries Department to promote the aquaculture system on a large scale. The Government provides 50 per cent subsidy assistance for the beneficiaries. So far, more than 4193 Jhora fishery units have been commissioned assisting under Fish Farmers Development Agency (FFDA) programmes. The depth of the Jhora is maintained 4 ft. in average. Common carp and Grass carp fingerlings introduced in the Jhora Fishery. After proper culture, the fish grow upto 500 gm. to 700 gm in a year. In this area temperature varies from 0° to 10° C and in this temperature region the growth is low, in comparison with plain area. Mortality rate is about 20 per cent. Fish Farmers Development Agency (FFDA) of Darjeeling introduced a model scheme of Jhora Fishery with some aims and objects. The main aims and object of the scheme was to help the poor fish farmers financially and to train them to culture fish in Jhora, as well as to assist them for Bank Loan. As fingerlings are not generally available in the hill areas of Darjeeling district, for maintaining proper fish culture selected fingerlings are collected from outside and distributed to the beneficiaries free of cost. A coldwater fish farm with seed production, rearing, table fish production and brood stock management facilities has already been constructed at Reyang.

The numerous big and small rivers, streams (Jhora) of Darjeeling district are the home of the important coldwater fish species and ornamental valued fish. But till today it has never been felt for sustainable exploitation of these precious fishes for income generation to the local people. This species is recommended for culture in artificial pools fed by perennial Jhora (Stream) are *Neolissocheilus hexagonalepis, Tor putitora, Schizothorax Richardson* along with exotic species, grass carp, silver carp, common carp and IMC. Already the farmers of hill areas are culturing these species along with exotic species resulted increased in production per unit area. Jhora Fishery is very much popular in the three hill sub-divisions of Darjeeling District to the common people. However, hill fishery development in sub-Himalayan West-Bengal should focus on aquaculture of suitable fish species, promoting sport fishing, development of farm/infrastructure and HRD activities.

With high dissolved oxygen and low nutrient levels, common carp and grass carp have been successfully established for Jhora fishery culture on supplementary diets wherein hill stream water is stored in small irregular type ditch by constructing barricade using earth, sand and stones etc. Often an area of 1500 sq. ft is termed as one unit Jhora. After proper culture, fish grow 200-300 g/yr at a temperature of 0-20 °C with 80 per cent survival. With a view to augment fish production and help poor farmers, concept of Jhora fishery was introduced in Darjeeling district during mid-eighties of the previous century and subsequently in the preceding years, Fish Farmers Development Agency (FFDA) came into existence for rendering assistance in obtaining bank loans, provide short term training with stipend and distribution of seed free of cost to the beneficiaries (Mukherjee, 2002). For quality fish seed self efficiency, State Fisheries Department in collaboration with Darjeeling Gorkha Hill Council (DGHC) undertook the construction of a fish seed farm at Rayang near Rambi bazar in Kalimpong sub-division in Darjeeling district (Barat *et al.*, 2005).

Along with exotic Chinese carps, indigenous food and sport fish-Chocolate mahseer (*Neolissocheilus hexagonolepis*) is considered as most preferred culturable coldwater fish species in Jhora ponds on account of easy adaptability and high market prices. Growth of katli could be as high as 250 g/6 months from an initial stocking size of 80-100 m.m. and fish upto 2 kg can be reared in these ponds (Barat *et al.*, 2005). Further, various aspects of breeding and rearing of this species are being carried out at fish seed farm at Rayang from the brood stock collected from nature. More plans are underway to set up a proper brood fish bank for different coldwater fish species in this farm.

2.5. Wetlands of North Bengal

(*i*) Status of Wetlands

Wetlands are one of the precious aquatic systems intimately associated with mankind and have been used extensively. They are the excellent repository of aquatic biodiversity besides many other functions, such as irrigation, potable water, recharger and decharger of ground water, flood protector and trappers of nutrients and toxicants. Wetlands of West Bengal covering more than 150 Beels and brackish water bheries constitute about 31 per cent of total freshwater resources of the state. They are related to rivers Bhagirathi, Hooghly, Icchamati, Jalangi, Churni, Kalindi, Dharub, Dharala, Pagla, Behula, Torsa and Purnabhaba and mainly spread over the Districts of North and South 24 Parganas, Hooghly, Murshidabad, Nadia, Maldah, South and North Dinajpur, Coochbihar etc.

West Bengal beels portray a relatively better picture of fish and fisheries as compared to other part of the country. The fishery is though mainly dominated by miscellaneous species, the major carps have indicated sizeable contribution. Catfish and live fishes are also common in total fish landings. The main commercially important species encountered were IMC. The percentage composition of fish catch in West Bengal Beels varies from district to district (Chakraborty, 2007a) depending on stocking, crafts and gear used, efficiency of fishers. The percentage composition of fish catch is as follows:

Table 4: Percentage Composition of Fish Catch in Beels of West Bengal

Group	*Per cent Composition*
IMC	40-45
Catfish	6-8
Murrells	7-10
Air-breathing	12-20
Miscellaneous	20-30

To attain self sufficiency in fish production is a challenging task and to achieve this goal, however, a scientific understanding of the ecological management of all these water bodies supporting capture fisheries is imperative to back up their optimum exploitation.

In open beels, the natural populations, their recruitment and growth are monitored to obtain the desired production levels whereas in the closed ones stocking is the mainstay of management. Many of these Beels have also become defunct and dead due to flood control measure like strengthening of the river embankments to protect low-lying areas. Most of the Beels require provision of sluice gates, desilting and deepening of the connecting channels and clearance of the dense weed infestation. There lies a great scope to reclaim many of the dead beels into productive fish farms.

'Culture based capture fishery' has been practiced in this state for last few years. The art of optimum exploitation of the beel fisheries revolves round the concept of keeping the deeper zones as exclusive areas for capture fisheries. The marginal areas are renovated for the development of culture fisheries. For such exercises the prime needs are desilting of the connecting channels and constructing the perimeter embankments. Towards this effort, Fisheries Department, West Bengal has taken up a reclamation project for the desiltation and renovation of beels in the different districts.

Now if these beel areas are desilted and renovated, the production from the beel fisheries may just be doubled, beside creating a significant impact in the conservation of ecological niches. And the overall impact will act upon in the socio-economic upliftment of the poor fishermen, wholly dependent on these beels significantly.

Wetlands are highly significant in terms of ecology, socio-economy and other benefits, showering blessings upon the surrounding human society. They also provide sustainable benefits that make a fair number of our population to earn their daily bread. In addition, a variety of life forms like water and migratory birds dwell there, which change the scenic beauty and produce wide range of valuable products. About 20 per cent of the material we use daily, especially food and drugs come directly from the wetlands, and most of the great human habitations are situated here.

(*ii*) SWOT Analysis

Wetlands in northern part of West Bengal are one of the prime and traditional fishery resources with tremendous scope for their fishery development, following scientifically based norms. Ecologically the wetlands are highly productive ecosystems, but reeling under various kind of stress factors. The production functions and biodiversity, necessary for human welfare, have developed aberrations. Public awareness and their active participation are necessary to conserve the dwindling resources, both physical as well as biological. In the light of this it becomes our prime duty to conserve the floodplain wetlands resources of this state. Its fishery development, following scientifically based norms may play a pivotal role in conserving the wetlands, increased fish production and gainful employment to thousand of people. The SWOT analysis for developing fisheries sector of wetlands as well as to conserve its physical entity in North Bengal is briefed as follows (Chakraborty, 2007a):

(*a*) Strength

- ☆ Wetlands are biologically sensitive habitats for economically important fish species
- ☆ Important fishery resources of the North Bengal
- ☆ Plays a vital role in the flood control
- ☆ Maintain agricultural irrigation
- ☆ Keep ecological balance
- ☆ Regulate the water regime and nutrient exchange and act as natural filters
- ☆ Support local peoples in different ways in their livelihood
- ☆ Managed by the Fisherman Co-operative Societies

(b) Weekness

- ☆ Highly threatened ecosystems and irrational fishing practices.
- ☆ Massive and thick stand of aquatic macrophytes
- ☆ Persistent organic pollutants in untreated waste water
- ☆ For long years of exploitation most of the large water bodies have become silted
- ☆ Lack of proper database of recent years
- ☆ Presence of different stakeholders
- ☆ Lack of knowledge of scientific management and conservation
- ☆ Lack of knowledge in functioning of Fishermen co-operative societies
- ☆ Poor socio-economic condition of users
- ☆ Financial institutions are hesitant to support
- ☆ Implementation of law and order

(c) Opportunities

- ☆ Conservation and sustainable utilization of biological resources
- ☆ Holistic approach to conserve the physical entity
- ☆ Maintain ecological balance essential for the life support system
- ☆ Restore biological damage caused by human interventions
- ☆ Developing suitable and effective management packages
- ☆ Development of culture base fisheries and management of capture fishery
- ☆ Integrating the fisheries development of wetlands
- ☆ Creation of awareness among the people about prestige status and conservation method of wetlands
- ☆ Increasing human participation in resource management
- ☆ Strengthening of Fishermen co-operative societies
- ☆ Fishermen training for better understanding of production functions

Various forms of enhancement (*viz.* Stock, Species, Environment and Management) therefore, hold the key for the sustainable fisheries development in these ecosystems. This development process will cause more production of fish, effective utilisation of water bodies for productive use and generation of more mandays.

(d) Threat

- ☆ High rate of siltation and catchment modifications
- ☆ Rampant encroachment for lands and human settlement
- ☆ Increasing incidents of eutrophication and pollution
- ☆ Increasing stands of unwanted biotic communities
- ☆ Over exploitation of resources for short term gains
- ☆ Irrational Leasing policy
- ☆ Inadequate financial support in time from financial institutions

- ☆ Poor functioning of co-operative societies
- ☆ Conflicts among various user groups
- ☆ Social tensions

For the development of the fisheries of wetlands, proper data base is very important. Base-line reliable data on resource inventory for planning and management for sustainable exploitation of wetland resources are urgently needed. The improper database is one of the weakest links in formulation of management policies for fisheries and aquaculture sectors of any area. Adequate and reliable database is required to be established, as it forms pre-requisite for plan of future course of action. Prior to build such a data base, it is emergent to standardize the nomenclature and classification for various relevant parameters, *e.g.* size of wetland, fish seed grades, groups of fish catch, infrastructure, fish based industry etc (Chakraborty, 2007a).

Thrust Areas for Wetlands Management

- ☆ Assessment and effective conservation of biodiversity and maintaining ecological elasticity
- ☆ Evolving quick evaluation techniques on environmental impact assessment
- ☆ Survey and mapping of wetland resources of the state
- ☆ Developing strategies for the control of siltation
- ☆ Effective control measures for prolific growth of unwanted biota and exotic species
- ☆ Developing eco-friendly and rational package of practices for fisheries development
- ☆ Monitoring to regulate the introduction of exotic fish species
- ☆ Integrated approach of management and diversification of activities
- ☆ Strengthen extension methods for increased productivity
- ☆ Upgrading knowledge of stakeholders on regular basis

In rural areas, there is a new awareness among the masses of the economic benefits that can accrue through improved pisciculture. Further progress will depend on the successful infusion of scientific eco-friendly technology and pragmatic managerial and commercial practices in breeding, rearing, harvesting, processing, storing and marketing operation in wetlands. Mechanism of transfer of technologies and extension functionaries are required to be strengthened with new dimensions. To save the physical entity of wetlands as well as to become self sufficient in fish production, initiative has to be taken to formulate an effective integrated management plan at the earliest with proper monitoring of activities in the frame work of environmental variables.

2.6. Constraints Faced in Aquaculture Process

1. Excess sediment particularly during the rainy season increase turbidity of water which results in poor penetration of light and reduces photosynthesis rate of phytoplankton and benthos resulting in decrease of food organism of the system.

2. Soil erosion following deforestation results in siltation covering spawning sites with sediments that can ultimately become a cause of loss of fish diversity.
3. Dams have been constructed over the different rivers. These dams act as physical barriers for migration, preventing access to fishes to their usual breeding, rearing and feeding grounds.
4. Lack of awareness about the importance of environmental conservation and the concept of sustainable fisheries development among the local fisher folk. Indiscriminate use of nylon gill nets of small mesh size, attempt to catch all the fishes in the dry season, to chase and catch fishes from their breeding ground –these practices cause over exploitation of the fishes and resulting in severe population disturbance.
5. Pesticides are applied in huge quantities in the tea plantation and the paddy fields. Chlorinated hydrocarbons, organophosphates and carbonates are some of the commonly used groups. These chemicals ultimately find their way into water systems, which cause abnormal behavior and increased mortality of fish.
6. Killing of fish by pesticides, copper sulphates, bleaching powder are rampant.

2.7. District-wise Aquatic Resources in North Bengal

District-wise fishery resources, production potentials and other relevant data of North Bengal are depicted in Tables 5–21.

Table 5: District-wise Impounded Water Area (In ha) in North Bengal

Sl.No.	*Name of the District*	*Culturable Area*	*Semi-derelict Area*	*Derelict Area*	*Total Area*
1.	Darjeeling	210.29	–	–	210.29
2.	Jalpaiguri	2404.53	3068.00	527.47	6000.0
3.	Coochbihar	567.86	990.76	327.44	1886.06
4.	Uttar Dinajpur	2910.80	978.00	816.00	4704.80
5.	Dakshin Dinajpur	5910.03	1056.00	1294.00	8260.03
6.	Malda	1958.32	3055.42	3557.23	8570.97
7.	Total	13961.83	9148.18	6522.14	29632.15
8.	West Bengal	211332.18	52087.89	24126.82	287546.89
9.	Per cent Share of North Bengal	6.61	17.56	27.03	10.31

The share of total pond and tank area of North Bengal in respect to West Bengal is 10.31 per cent, of which culturable area is 6.61 per cent, semi-derelict area is 17.56 per cent and derelict area is 27.03 per cent.

Table 6: District-wise Area (in ha) of River, Canal/Khal, Beel and Baor in North Bengal

Sl.No.	Name of the District	River	Canal/Khal	Beel/Baor	Total Area
1.	Siliguri	3395.63	683.94	683.94	4763.51
2.	Jalpaiguri	16006.44	562.29	504.38	17.073.11
3.	Coochbihar	13095.82	1664.58	1168.11	15928.51
4.	Uttar Dinajpur/ Dakshin Dinajpur	4032.08	1774.30	3548.59	9345.97
6.	Malda	6497.6	2511.42	4551.55	13560.57
7.	Total	43027.57	7196.53	10456.57	43615.743
8.	West Bengal	164525.99	80085.71	42085.71	286693.35
9.	Per cent Share of North Bengal	26.15	8.99	24.85	15.21

The share of total water resource (in ha) of River, Canal/Khal, Beel and Baor of North Bengal in respect to West Bengal is 15.21 per cent, of which river area is 26.15 per cent, canal/khal area is 8.99 per cent and beel/boar area is 24.85 per cent.

Table 7: Number of Hatcheries in North Bengal

Sl.No.	District	IMC Hatchery (Govt)	IMC Hatchery (Private)	Prawn Hatchery	Ornamental Hatchery
1.	Darjeeling		0		
2.	Jalpaiguri		7		2
3.	Cooch Bihar		15		
4.	Uttar Dinajpur	1	17		
5.	Dakshin Dinajpur		11		
6.	Maldah	1	7		
7.	Total	2	57	0	2
8.	West Bengal	6	588	–	4
9.	Per cent Share of North Bengal	33.33	9.69		50.00

The share of no. of Hatcheries present in North Bengal in respect to West Bengal is 9.69 per cent (Private Hatchery) and 33.33 per cent (Govt. Hatchery) respectively.

The share of Fish seed production of North Bengal in respect to West Bengal has increased from 3.09 per cent during 2002-03 to 6.32 per cent during 2008-09.

The share of total Inland fish production of North Bengal in respect to West Bengal during 2005-06 to 2008-09 varies from 11.22 per cent to 11.76 per cent in case of fish production.

Table 8: District-wise Seed Production (in million) in North Bengal

Sl.No.	*Name of the District*	*02-03*	*03-04*	*04-05*	*05-06*	*06-07*	*07-08*	*08-09*
1.	Coochbehar	25	34	37	40	82	90	38
2.	Jalpaiguri	30	42	50	70	87	125	145
3.	Darjeeling (Siliguri)	0	0	0	0	0	0	0
4.	Uttar Dinajpur	20	25	30	50	113	81	101
5.	Dakshin Dinajpur	130	140	150	200	377	336	420
6.	Malda	150	200	210	250	34	172	181
7.	Total:	355	441	477	610	693	804	885
8.	West Bengal	9100	11890	12200	13160	13200	13572	14000
9.	Per cent Share of North Bengal	3.90	3.71	3.91	4.64	5.25	5.92	6.32

Table 9: District-wise Inland Fish Production (in ton) in North Bengal

Sl.No.	*District*	*2005-06*		*2006-07*		*2007-08*		*2008-09*	
		Fish	*Prawn*	*Fish*	*Prawn*	*Fish*	*Prawn*	*Fish*	*Prawn*
1.	Darjeeling	342	–	342	–	352	0	330	0
2.	Jalpaiguri	15000	–	15000	–	23456	0	28505	0
3.	Cooch Bihar	19090	–	19090	–	21257	0	21165	0
4.	Uttar Dinajpur	14829	–	14829	–	17225	0	18813	0
5.	Dakshin Dinajpur	32464	–	32464	–	35455	0	36528	0
6.	Malda	38116	10	38116	10	36564	7	38420	2
7.	Total	119841	10	119841	10	134309	7	143761	2
8.	West Bengal	1028 903	61100. 25	1028 903	61100. 25	1196 874	67570	1222 510	72200
9.	Per cent Share of North Bengal	11.65	0.02	11.65	0.02	11.22	0.01	11.76	0.00

Table 10: District-wise and Year-wise Water Area (in ha) Covered Under FFDA in North Bengal

Sl.No	*District*	*2001-02*	*2002-03*	*2003-04*	*2004-05*	*2005-06*
1.	Uttar Dinajpur	53.450	67.900	68.440	80.210	76.810
2.	Dakshin Dinajpur	50.480	60.080	40.650	45.220	46.490
3.	Siliguri	5.430	5.480	1.330	5.460	5.110
4.	Jalpaiguri	85.560	64.070	60.070	65.160	54.620
5.	Coochbihar	20.460	44.470	47.38	19.480	48.210
6.	Malda	41.460	38.930	54.15	57.360	48.170

Contd...

Table 10–Contd...

Sl.No	District	2001-02	2002-03	2003-04	2004-05	2005-06
7.	Total	256.84	280.93	272.02	272.89	279.41
8.	West Bengal	2325.990	2110.100	1912.99	1700.410	1087.36
9.	Per cent Share of North Bengal	11.04	13.31	14.22	16.05	25.70

Table 11: District-wise and Year-wise Water Area (in ha) Covered (Cumulative) Under FFDA

Sl.No.	District	2006-07	2007-08	2008-09	Total (Cumulative)
1.	Uttar Dinajpur	71.06	64.83	51.73	1014.51
2.	Dakshin Dinajpur	47.34	47.22	25.31	5660.06
3.	Siliguri	5.93	8.22	2.69	199.45
4.	Jalpaiguri	62.7	30.05	25.46	2237.53
5.	Coochbihar	65.26	51.78	41.52	2103.342
6.	Malda	44.27	60.3	48.49	3867.898
7.	Total	296.56	262.4	195.2	15185.54
8.	West Bengal	1625.31	1585.07	1233.95	123817.7
9.	Per cent Share of North Bengal	18.25	16.55	15.82	12.26

The share of total water area (in ha) covered (cumulative) under FFDA of North Bengal in respect to West Bengal is 12.26 per cent.

Table 12: F.F.D.A Production (in ton) in North Bengal

Sl.No.	District	2004-05	2005-06	2006-07	2007-08	2008-09
1.	Coochbihar	7206.14	7720.78	8582.88	8968.93	9254.7
2.	Jalpaiguri	7514.51	8413.70	9317.22	9622.5	9845.13
3.	Darjeeling	385.31	407.92	415.8	436.4	466.07
4.	Siliguri	909.15	724.96	805.06	855.91	877.58
5.	Uttar Dinajpur	2674.98	3282.75	3834.25	4188.09	4463.84
6.	Dakshin Dinajpur	20586.80	21994.55	23858.75	24511.16	24904.26
7.	Malda	13665.92	14747.91	16051.39	16614.42	17018.75
8.	Total	52942.81	57292.57	62865.35	65197.4	66830.3
9.	West Bengal	441849.43	473912.28	516641.59	533229	544812
10.	Per cent Share of North Bengal	11.98	12.09	12.17	12.23	12.27

The per cent share of FFDA production of North Bengal in respect to West Bengal is increased from 11.98 per cent to 12.27 per cent from 2004-05 to 2008-09.

Table 13: District-wise No. of Beneficiaries upto 2008-09 Since Inception Under FFDA in North Bengal

Sl.No.	*District*	*No. of Beneficiaries Under FFDA*
1.	Darjeeling (DGHC)	5040
2.	Darjeeling (Siliguri)	860
3.	Coochbihar	6414
4.	Jalpaiguri	15815
5.	Uttar Dinajpur	3811
6.	Dakshin Dinajpur	12459
7.	Malda	16956
8.	Total	61355
9.	West Bengal	402185
10.	Per cent Share of North Bengal	15.26

The share of the no. of beneficiaries under FFDA of North Bengal in respect to West Bengal is 15.26 per cent during 2008-09.

Table 14: Fish Based Industries in North Bengal

Sl.No.	*Name of the District*	*Ice Plant*		*Feed Plant*		*Fish Drying Centre*	
		No	*Capacity (t)*	*No*	*Capacity (t)*	*No*	*Capacity (t)*
1.	Darjeeling	5	380.8	–	–	–	–
2.	Jalpaiguri	4	22.2	–	–	–	–
3.	Cooch Bihar	–	–	–	–	–	–
4.	Uttar Dinajpur	20	16.5	1	27.5	–	–
5.	Dakshin Dinajpur	13	20.7	–	–	–	–
6.	Malda	4	37			15	131.1
7.	Total	46	477.2	1	27.5	15	131.1
8.	West Bengal	288	5241	12	99	352	12923.7
9.	Per cent Share of North Bengal	15.97	9.11	8.33	27.78	4.26	1.01

The per cent share of number of Ice Plant is 15.97 per cent (Capacity-9.11 per cent), Feed Plant is 8.33 per cent (Capacity-27.78 per cent) and Fish drying centre is 4.26 per cent (Capacity-1.01 per cent) in respect to West Bengal.

Table 15: Carp Hatchery Plant, *Macrobrachium* Hatchery and Demonstration Centre in North Bengal

Sl.No.	*District*	*Carp Hatchery Plant*	*Macrobrachium Hatchery*	*Demonstration Centre*
1.	Darjeeling	1	0	5
2.	Jalpaiguri	7	0	0
3.	Cooch Bihar	15	0	0
4.	Uttar Dinajpur	16	0	1
5.	Dakshin Dinajpur	7	0	0
6.	Malda	4	0	7
7.	Total	50	0	13
8.	West Bengal	350	10	189
9.	Per cent Share of North Bengal	14.29	0.00	6.88

The per cent share of North Bengal in respect to West Bengal in Carp Hatchery Plant, and Demonstration Centre is 14.29 per cent, and 6.88 per cent respectively. No *Macrobrachium* Hatchery is established till date.

Table 16: Wholesale and Retail Fish Market in North Bengal

Sl.No.	*Name of the District*	*Market*			
		Whole Sale		*Retail*	
		No.	*No. of Wholesalers*	*No.*	*No. of Retailers*
1.	Darjeeling	1	140	33	975
2.	Jalpaiguri	16	69	161	2942
3.	Cooch Bihar	52	270	148	2470
4.	Uttar Dinajpur	16	105	31	800
5.	Dakshin Dinajpur	8	94	51	756
6.	Malda	16	240	95	1498
7.	Total	109	918	519	9441
8.	West Bengal	408	4115	2882	35988
9.	Per cent Share of North Bengal	26.72	22.31	18.01	26.23

The share of Whole sale markets of North Bengal is 26.72 per cent, where the no. of Wholesalers is 22.31 per cent and for retail market it is 18.01 per cent with 26.23 per cent Retailers.

Table 17: District-wise Fisher Folk Population in North Bengal

Sl No.	Name of District	No. of Fishing Village	No. of Fisher Folk Family	Fisher folk population		
				Male	Female	Total
1.	Darjeeling	5	956	1677	1282	2959
2.	Jalpaiguri	282	33039	92807	51503	144310
3.	Coochbihar	643	22538	71911	71277	143188
4.	Uttar Dinajpur	221	11925	40447	29903	70350
5.	Dakshin Dinajpur	219	16601	43691	26262	69953
6.	Malda	55	13857	46347	42327	88675
7.	Total	1425	98916	296880	222554	519435
8.	West Bengal	6088	535727	1654238	1283943	2938181
9.	Per cent Share of North Bengal	23.41	18.46	17.95	17.33	17.68

In respect to West Bengal the per cent share of number of Fisher folk family of North Bengal is 18.46 per cent. Fisher folk population is 17.68 per cent of the whole state.

Table 18: District-wise Numbers of Family Members Engaged in Fishery Activities in North Bengal

Sl.No.	Name of the District	Fisher Folk Fulltime			Fisher Folk Part-time			Fisher Folk Occasional		
		Male	Female	Total	Male	Female	Total	Male	Female	Total
1.	Darjeeling	391	57	448	552	74	626	734	1151	1885
2.	Jalpaiguri	11128	430	11558	8050	780	8830	1035	90	1125
3.	Coochbihar	5617	872	6489	8943	2985	11928	1806	218	2024
4.	Uttar Dinajpur	7291	–	7291	18226	–	18226	10935	–	10935
5.	Dakshin Dinajpur	7406	3539	10945	5563	2881	8444	6125	4704	10829
6.	Malda	23174	2116	25290	6952	847	7799	2317	423	2740
7.	Total	55007	5655	62021	48286	3728	55853	22952	5127	29538
8.	West Bengal	496004	137619	633623	431793	225212	657004	265063	126235	391298
9.	Per cent Share of North Bengal	11.09	4.11	9.79	11.18	1.66	8.50	8.66	4.06	7.55

The per cent share of North Bengal of fulltime fisher folk families is 9.79 per cent (male-11.09 per cent, female-4.11 per cent), part-time Fisher folk is 8.50 per cent (male-11.185, female-1.66 per cent) and occasional Fisher folk is 7.55 per cent (male-8.50 per cent, female-4.06 per cent) in respect to West Bengal.

Table 19: District-wise No. of Co-operative Societies of North Bengal upto 2008–09

Sl.No.	Districts	No. of Central Fishermen Cooperative Societies (CFCS)	Primary Fishermen Co-operative Society (PFCS)				
			Category A	Category B	Category C	Total	No. of Ornamental Co-operative Societies
1.	Darjeeling	1	1	0	1	2	0
2.	Jalpaiguri	1	15	0	2	17	1
3.	Cooch Bihar	1	57	4	0	61	0
4.	Uttar Dinajpur	1	14	19	0	33	11
5.	Dakshin Dinajpur	1	13	4	0	17	8
6.	Malda	1	81	2	0	83	13
7.	Total	6	181	29	3	213	33
8.	West Bengal	20	489	157	401	1047	125
9.	Per cent Share of North Bengal	30.00	37.01	18.47	0.75	20.34	26.40

The per cent share of CFCS in North Bengal is 30.00 per cent in respect to West Bengal. The per cent of total no. of PFCS is 20.34 per cent and the no. of Ornamental Co-operative societies is 26.40 per cent of the state.

Table 20: District-wise Employment Generation in Fisheries Sector of North Bengal

Sl.No.	Districts	Inland Fisheries	Extension Wing	Welfare Scheme	Co-op Society	Others	Total Man-days	Total Man-year
1.	Darjeeling (DGHC)	5900	472	120	2370	1145	10007	42
2.	Darjeeling (Siliguri)	1629	330	30		45	2034	8
3.	Jalpaiguri	103271	1029	635	3671	1645	110251	459
4.	Cooch Bihar	308	4912	2184	7280	1345	16029	67
5.	Uttar Dinajpur	49787	2592	10	27090	63	79542	331
6.	Dakshin Dinajpur	8400	1170	1075	35580	2976	49201	205
7.	Malda	48507	1155	280	115770	660	166372	693
8.	Total	217802	11660	4334	189391	7879	433436	1805
9.	West Bengal	3126687	55867	140987	3024903	234226	17847560	74365
10.	Per cent Share of North Bengal	6.97	20.87	3.07	6.26	3.36	2.43	2.43

The per cent shares of North Bengal in employment generation in Inland fisheries, Extension wing, Welfare scheme, Co-op society and others during 2005-06 are 6.97 per cent, 20.87 per cent, 3.07 per cent, 6.26 per cent and 3.36 per cent respectively.

Whereas the per cent shares of total man-days is 2.43 per cent and of total man-years is 2.43 per cent in respect to whole state.

Table 21: District-wise Short Term Credit Programme in Fisheries Sector of North Bengal

Sl.No.	Name of the District	2006-07		2007-08	
		Area Involved (ha)	No. of Beneficiaries	Area Involved (ha)	No. of Beneficiaries
1.	Darjeeling (DGHC)	62 JU	60	0	0
2.	Darjeeling(Siliguri)	6.31	21	8.38	11
3.	Coochbihar	56.27	42	96.74	26
4.	Jalpaiguri	68.20	163	0	0
5.	Uttar Dinajpur	11	18	0	0
6.	Dakshin Dinajpur	53.31	35	44.69	43
7.	Malda	22.02	5	51.25	34
8.	Total:	217.11	344	201.06	114
9.	West Bengal	3380.92	4315	2695.1	2920
10.	Per cent Share of North Bengal	6.42	7.97	7.46	3.90

The per cent share of North Bengal of area involved in short term credit programmes is ranging from 6.42 per cent to 7.46 per cent during 2007-2008.

2.8. District-wise Rivers and Big Water Bodies of North Bengal

The name of rivers (Figures 2 and 3) of North Bengal and their length are given below:

(*i*) Darjeeling District

(a) Rivers and Big Water Bodies of Darjeeling District

Darjeeling the 'Queen of Hills' in North Bengal is the most popular tourist destination in northeast India. The district is situated at an altitude of 2134 m. Darjeeling with its exquisite scenic beauty attracts travellers of varied interest. Darjeeling is the Abode of Lord Shiva offering spectacular views of snowy peaks of Kanchanjunga and the hospitality of colorful Himalayan people. Darjeeling town (area, 11.44 km^2) is located on a spur with houses hugging the hillside.

Most probably the name Darjeeling has come from the words 'Dorgiling' or 'Durjayling'. The district Darjeeling is on the head of West Bengal. As the district is located at the south of Himalaya, every river (Figures 4) flows down towards south from north. Every rivulets or a small river of the district flows to the two main rivers – Teesta and Mahananda. Darjeeling district is hilly and woody. It is rich in wild life. Its area is about 3149 sq.km. The rivers of the district are (Plate 2-9) as follows.

Table 22: Length of Rivers

Sl.No.	Name of Rivers	Length (KM)	Sl.No.	Name of Rivers	Length (KM)
1.	Teesta	37.00	2.	Rangeet	42.34
3.	Rishi	17.36	4.	Reyang	18.70
5.	Relli	NA	6.	Jaldhaka	19.47
7.	Gail Khola	NA	8.	Pala	NA
9.	Raman Khola	NA	10.	Balasun	48.40
11.	Lodama Khola	NA	12.	Shir Khola	NA
13.	Leesh	12.10	14.	Ghish	30.20
15.	Chel	10.46	16.	Mahananda	91.70
17.	Raidak	NA	18.	Sankosh	NA
19.	Torsha	NA	20.	Rungdung Khola	39.78
21.	Dudia Khola	NA	22.	Rongtong Khola	NA
23.	Rohini	NA	24	Tarkhola	NA
25.	Neem	NA	26.	Geet	NA
27.	Nimbong	NA	28.	Newra	NA

Mechi

Mechi-chowki, the hills of the eastern border of Nepal is the source of this river. Flowing down towards south, it passes to Purnia district of Bihar near Naxalbari. Here one river named Nanda from Nepal falls in Mechi on its left side. From this place of unison, Burimechi flows towards southwest.

Balasun

The name Balasun is due to the golden colour of sand of this river. Originally from Lepajagat, the river meets Bheem khola, Pulangdong, Rangmukh west drain as it flows down. Besides this, Rangbang (right tributary) and Ringchingtong khola (left tributary) falls at Balasun near Mirik. Rakvi and Rohini rivers originating from Kurseong hill meet Balasun flowing forward by west of Siliguri fall in Mahanadi (Mahananda). Following are its worth mentioning tributaries:

Manjadhara and Changnala

Both originate near Khorabari.

Rangtang Nala

Originating from the Tindharia, flowing down in a zig zag way, it meets Balasun near Mahananda-Balasun junction. This is a left tributary.

Pancholan River

Its born place is also Tindharia. Flowing from the south of Tindharia it meets Balasun near Mahananda-Balasun junction.

Mahanadi (Mahananda)

The name Mahanadi comes from the Lepcha word 'Mahaldi'. The meaning of

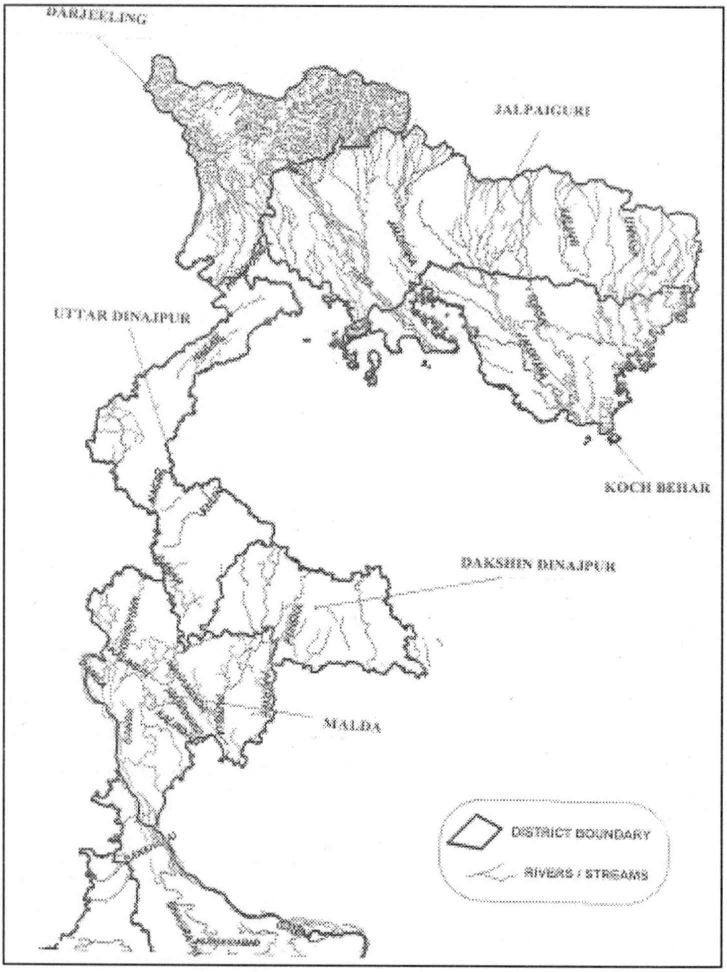

Figure 2: River Map of North Bengal

Mahaldi is bent. In Mahaldiram, on the north of Kurseong, a number of nameless streams coming together and give birth to Mahanadı (Mahaldi). Its length is about 290 km. Leaving Darjeeling district; Mahananda crosses Bihar, Bangladesh and falls in the river Padma.

Teesta

This is a very significant river of North Bengal. There are many myths on Teesta. 'Teesta' is a name of ancient village. Probably it comes from word 'Distang'. Many say that modification of the word 'Trisrota' is Teesta. Here 'Trisrota' meaning tri-flow refers to three famous rivers Kortoa, Atria and Punarvaba of North Bengal. Niharranjan Roy says–'The river Kortoa has entered Bangladesh passing through Darjeeling and Jalpaiguri district and originating from Bhutan border and north Himalaya. This northward flow gives the name Teesta. Passing Jalpaiguri it is divided into three flows – southward of Kortoa, middle one in Atrai and the westward in

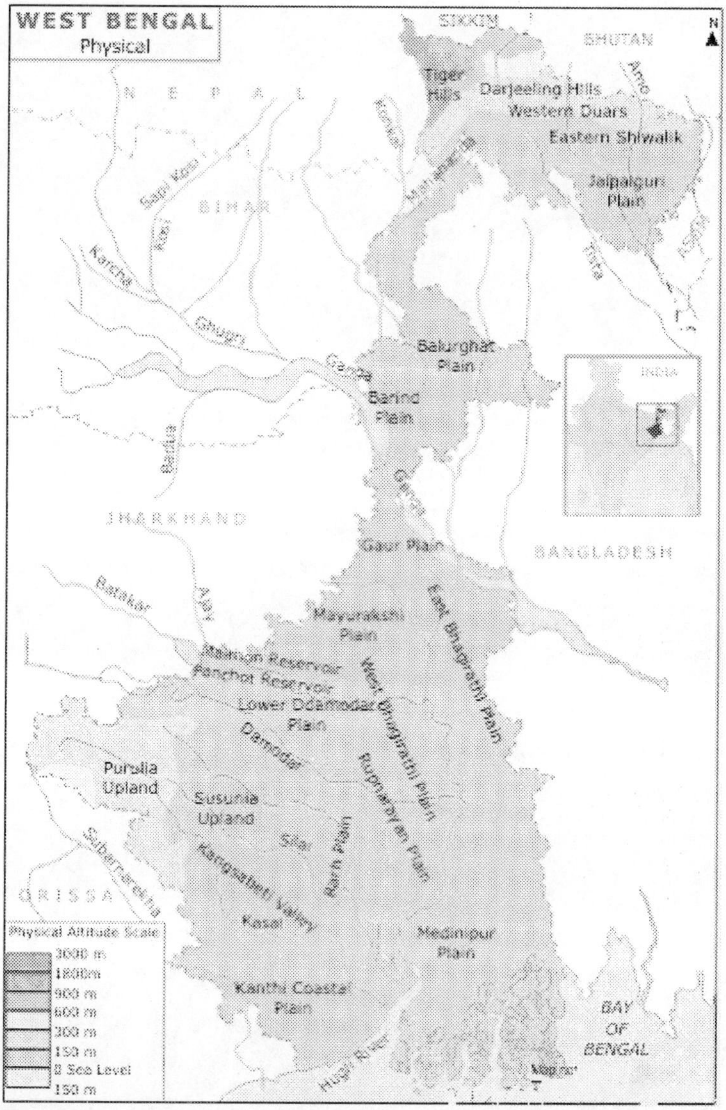

Source: www.mapsofindia.com

Figure 3: River Map of West Bengal

Punarvaba. At present Teesta does not flow as tri-flow. In the north of Sikkim, Teesta has originated from an iceberg called Teesta-Thangse (Height 6400 mt). It enters West Bengal touching the border city of west Sikkim Rangpo. Teesta meets Rongo nala in this Rangpo. In the district of Jalpaiguri there are a lot of tributaries that meet Teesta. Some of them are as follows.

Rango Nala

Falls in Teesta passing by Mongpu. This is a Right tributary.

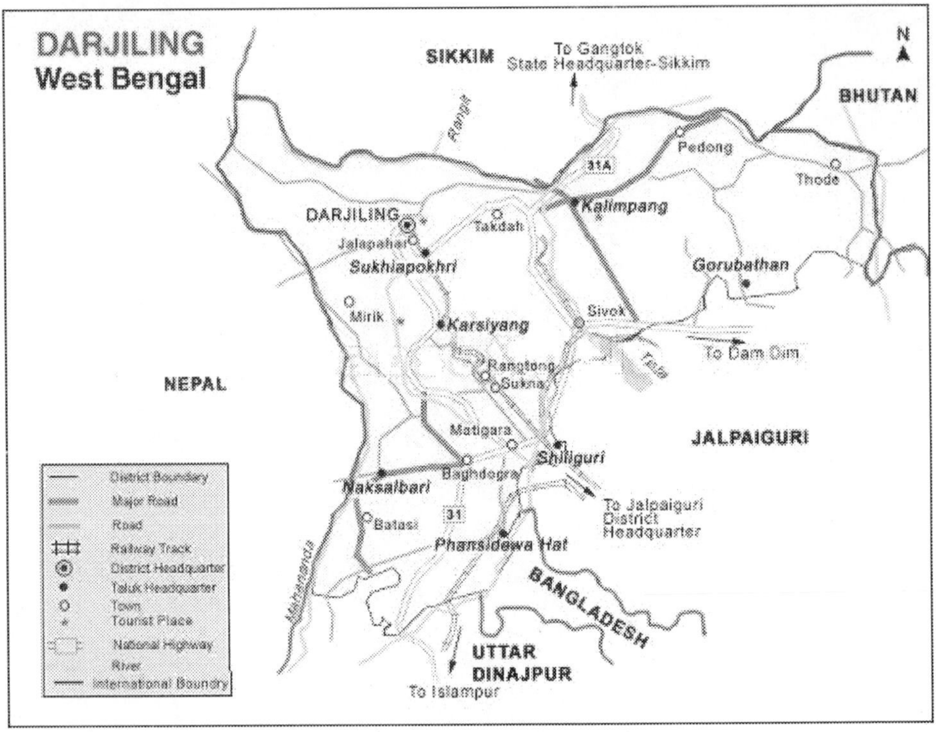

Source: www.mapsofindia.com

Figure 4: River Map of Darjeeling District

Reyang Nala

Born in Mahaldiram and falls in Teesta flowing northward. This is a Right tributary.

Lish Nala

Born in Lule village, passing by Bagracot tea garden falls in Teesta. It has been arised mixing Turang canal and Lish canal.

Gish Nala

Born in the south of the Kalimpong, flowing down towards south, it goes to Jalpaiguri district. Nearby the source of Gish, Lethi and Ramthi has also originated and mixed with Gish.

Chel Nala

The back of Gitbiyang is the source of this river. It flows down towards south mixing with Sukha and Malejing rivers.

Sitikhola

Arising from forests of Latpanchar flowing towards right side of Teesta.

Plate 2: Confluence of River Teesta and Rangeet–One of the Best Angling Spot

Plate 3: Aerial View of Mighty River Teesta

Plate 4: River Mahananda

Plate 5: River Balason

Plate 6: River Torsa

Plate 7: River Buribalason

Plate 8: River Jarda

Plate 9: River Jaldhaka

Kalijhora

It arises from Latpanchar and falls in Teesta near Kalijhora village. This is a Right tributary.

Sebak Khola

Its born place is hilly forest of Latpanchar. Meet Teesta near Sevoke. This is a right tributary.

Rellikhola

Born in Kalimpong hills, almost near the border of Bhutan and Darjeeling on the east of Kalimpong hill, it flows down towards south. Later flowing towards southwest it meets Teesta near Teesta bazaar. It is a left tributary of Teesta.

Neora Nala

Arising from Kalimpong hill, flowing towards Jalpaiguri, meets a lots of river. Later falls in Teesta along with Karala river.

The largest tributary of Teesta in the district of Darjeeling is Rangeet.

Two main tributary of the river is – Ramam and Choto Rangeet.

Bara Rangeet

Its born place is in the hills near Gezing village of Sikkim. Near the place Raman and Choto Rangeet united with Bara Rangeet (Rangeet). Later, flowing through Darjeeling district, by Teesta market it falls in Teesta. Rangeet is the largest tributary of Teesta. This Right tributary is often called as Bara Rangeet. The water of Teesta is turbid while the water of Rangeet is transparent.

Jaldhaka

Arising from the snow covered hilly areas of Sikkim, the river flows down by the border of Bhutan and Darjeeling and falls in Teesta in Jalpaiguri. Dichu, Lichu, Para, Langchu, Rangchu and Machu etc. rivulets give birth to Jaldhaka.

The block-wise water resources in upland areas of Darjeeling district are provided in Tables 23–32.

Table 23: Estimated Total Water Area in Upland Areas of Darjeeling District

Sl.No.	Name of Blocks	Estimated Total Water Area (ha)
1.	Darjeeling Pulbazar	76.49
2.	Rangli Rangliot	43.79
3.	Kalimpong -I	49.55
4.	Kalimpong–II	65.63
5.	Gorubathan	19.96
6.	Jorebunglow Sukiapokhri	52.00
7.	Mirik	12.04
8.	Kurseong	18.36
	Total	337.82

Table 24: Detailed Distribution of Water Area in Darjeeling Phulbazar Block

Sl.No.	*Name of Village*	*Estimated Total Water Area ha*
1.	Rimbick	4.50
2.	Namla	0.67
3.	Lodhama	1.63
4.	Dangia	1.23
5.	Majua	0.78
6.	Kankibong	2.24
7.	Jhepi	1.13
8.	Lamagaon	1.59
9.	Relling	2.31
10.	Kaijalia	2.66
11.	Samalbong	1.34
12.	Kolbong	0.48
13.	Murmidong	1.17
14.	Karmi	0.58
15.	Goke	5.35
16.	Bijanbari	4.01
17.	Chongtong Tea Garden	4.22
18.	Liza Hill Tea Garden	1.18
19.	Singla Tea Garden	2.25
20.	Badamtam Tea Garden	3.87
21.	Ging Tea Garden	3.03
22.	Phubsering Tea Garden	2.25
23.	Rungneet Tea Garden	0.99
24.	Tukvar Tea Garden	3.12
25.	Lebong Tea Garden	1.68
26.	Soom Tea Garden	2.62
27.	Singtam Tea Garden	4.50
28.	Happy Valley Tea Garden	0.94
29.	Rishihat Tea Garden	1.44
30.	Rishihat Khasmahal	0.60
31.	Darjiling (P)	1.64
32.	Alubari Tea Garden	0.30
33.	Lebong and Mineral Spring Tea Garden	4.05
34.	Pandam Tea Garden	3.09
35.	Alubari Basty	0.33
36.	Pattabong Tea Garden (CT)	1.36
37.	Pattabong Tea Garden (CT)–Ward No.1	1.36
	Total	76.49

Table 25: Detailed Distribution of Water Area in Rangli Rangliot Block

Sl.No.	*Name of Village*	*Estimated Total Water Area ha*
1.	Kambal Tea Garden	4.59
2.	Lapchu Tea Garden	2.29
3.	Bara Mungwa	1.61
4.	Chhota Mungwa	0.87
5.	Takling Khasmahal	1.43
6.	Soriang Khasmahal	1.92
7.	Kolbong Khasmahal	0.73
8.	Rayak Khasmahal	0.12
9.	Lingding Khasmahal	0.98
10.	Hum Tukdah Khasmahal	3.17
11.	Tukdah Forest	1.94
12.	Lapchu Khasmahal	1.03
13.	Singringtam	1.07
14.	Tukdah	3.95
15.	Tukdh Tea Garden	4.17
16.	Rangli Rangliot Tea Garden	3.33
17.	Teesta Valley Tea Garden	7.15
18.	Reshop Bazar	1.02
19.	Labda Khasmahal	1.79
20.	Rong Chong Khasmahal	0.63
	Total	43.79

Table 26: Detailed Distribution of Water Area in Kalimpong–I Block

Sl.No.	*Name of Village*	*Estimated Total Water Area ha*
1.	Homes St.And Graihms	3.20
2.	Kaffir Khasmahal	0.88
3.	Kanke Bong Khasmahal	0.78
4.	Icha Khasmahal	3.04
5.	Pudung Khasmahal	1.91
6.	Sindibong Khasmahal	3.41
7.	Dungra Khasmahal	5.30
8.	Bhalukhop Khasmahal	4.74
9.	Kalimpong Khasmahal	6.70
10.	Bong Khasmahal	3.02
11.	Yokprintam Khasmahal	0.69
12.	Slokbhir Khasmahal	0.86

Contd...

Table 26–Contd...

Sl.No.	*Name of Village*	*Estimated Total Water Area ha*
13.	Samalbong Khasmahal	1.56
14.	Lulagaon Khasmahal	0.74
15.	Pemling Khasmahal	1.42
16.	Nimbong Khasmahal	2.13
17.	Nobgaon Khasmahal	1.29
18.	Paringar Khasmahal	1.43
19.	Samether Khasmahal	1.71
20.	Singi Khasmahal	1.06
21.	Riayang Railway Station	0.23
22.	Suruk Khasmahal	1.26
23.	Yang Makum Khasmahal	2.19
	Total	49.55

Table 27: Detailed Distribution of Water Area in Kalimpong–II (Algarah) Block

Sl.No.	*Name of Village*	*Estimated Total Water Area ha*
1.	Mansong Cinchona Plantation	9.62
2.	Kashone Khasmahal	2.69
3.	Pedong Bazar D.I.F.	1.06
4.	Pedong Khasmahal	6.18
5.	Kagey Khasmahal	1.81
6.	Maria Khasmahal	1.21
7.	Ladam Khasmahal	3.08
8.	Lingsaykha Khasmahal	2.03
9.	Lingsay Khasmahal	2.77
10.	Lava Bazar D.I.F.	1.82
11.	Sakiyong Khasmahal	5.89
12.	Sangser Khasmahal	3.76
13.	Dalapachan Khasmahal	3.97
14.	Paiyong Khasmahal	5.76
15.	Santuk Khasmahal	3.31
16.	Git Beong Khasmahal	2.41
17.	Git Dubling Khasmahal	3.63
18.	Paygang Khasmahal	0.71
19.	Bokhim Khasmahal	0.02
20.	Palla Khasmahal	1.89
21.	Lolay Khasmahal	2.01
	Total	65.63

Table 28: Detailed Distribution of Water Area in Gorubathan Block

Sl.No.	Name of Village	Estimated Total Water Area ha
1.	Rechila Chak Khasmahal	0.00
2.	Tuday Tangta Khasmahal	1.37
3.	Paten Godak Khasmahal	5.47
4.	Samsing Khasmahal	1.80
5.	Kumai Khasmahal	1.21
6.	Pankhasari Khasmahal	2.42
7.	Pashiting Khasmahal	0.37
8.	Nim Khasmahal	0.87
9.	Pagrang Bong Khasmahal	1.69
10.	Manabari Khasmahal	0.00
11.	Turibari Khasmahal	0.00
12.	Dalingma Khasmahal	0.48
13.	Gorubathan Khasmahal	3.45
14.	Mal Khasmahal	0.83
	Total	19.96

Table 29: Detailed Distribution of Water Area in Jorebunglow Sukiapokhri Block

Sl.No.	Name of Village	Estimated Total Water Area ha
1.	Salu Tea Garden	0.44
2.	Mariabong Tea Garden	1.29
3.	Lingia Tea Garden	1.20
4.	Tumsong Tea Garden	1.05
5.	Tumsong Khasmahal	0.34
6.	Mim Tea Garden	1.11
7.	Parmaguri Khasmahal	0.64
8.	Pulung Dong Khasmahal	1.29
9.	Pulung Dong Tea Estate	0.34
10.	Pubong Tea Garden	0.89
11.	Pussinbong Tea Garden	1.65
12.	Rang Bul	2.11
13.	Sonada Khasmahal	5.87
14.	Kalej Valley Tea Garden	1.56
15.	Dooteria Tea Garden	3.13
16.	Rongmook Ceder Tea Garden	3.25

Contd...

Table 29–Contd...

Sl.No.	Name of Village	Estimated Total Water Area ha
17.	Moonda Kotee Tea Garden	2.25
18.	Nahori Tea Garden (Balasom)	1.59
19.	Sukhiapokhri	2.56
20.	Achhalal Hatta (Rong Bong Basty)	1.71
21.	Soolbongs Tea Garden	1.39
22.	Seyok Tea Garden	1.25
23.	Gopal Dhara Tea Garden	1.18
24.	Mangarjung Tea Garden (Nagri)	2.96
25.	Sagmaru Tea Garden (Sungma)	1.39
26.	Tarzun Tea Garden	1.34
27.	Pokhribong Khasmahal	1.57
28.	Samrik Tea Garden (Simripani)	1.06
29.	Molatey Tea Garden (Aveneroge)	1.40
30.	Dhajea Tea Garden	0.89
31.	Nagri Farm Tea Garden	2.32
32.	Dhajea Khasmahal	0.98
	Total	52.00

Table 30: Detailed Distribution of Water Area in Mirik Block

Sl.No.	Name of Village	Estimated Total Water Area ha
1.	Okaity Tea Garden	1.09
2.	Bukim Tea Garden (Tharbu) (P)	1.64
3.	Phuguri Forest	0.02
4.	Paschim Phuguri Tea Garden	0.80
5.	Saurinibasti	1.46
6.	Mirik Khasmahal (P)	1.46
7.	Marma Tea Garden	1.15
8.	Singbulli Tea Garden	1.14
9.	Purba Phuguri Tea Garden	0.39
10.	Tingling Tea Garden	0.53
11.	New Fallodi Tea Garden (Ghyabari)	1.57
12.	Patong Tea Garden	0.79
	Total	12.04

Table 31: Detailed Distribution of Water Area in Kurseong Block

Sl.No.	*Name of Village*	*Estimated Total Water Area ha*
1.	Maharani Tea Garden	0.39
2.	Knaria Busty	0.26
3.	Singel Tea Garden	0.99
4.	Montiviot Tea Garden	0.30
5.	Karbia Tea Garden	0.38
6.	Castleton Tea Garden	0.40
7.	Spring Sidi Tea Garden	0.28
8.	Ambootia Tea Garden	1.81
9.	Makai Bari Tea Garden	0.65
10.	Longview Tea Garden	1.79
11.	Garidhora (M.Bari)Tea Garden	1.17
12.	Saptiguri Khelaghar Tea Garden	0.19
13.	Pairikumari	0.20
14.	Simulbari Tea Garden	1.31
15.	Rohini Tea Garden	0.89
16.	South Shibkhola Tea Garden	0.14
17.	Nurbang Tea Garden	0.38
18.	Mahanadi Tea Garden	0.32
19.	North Shibkhola Tea Garden	0.47
20.	Gidda Pahar	0.88
21.	St.Marys	0.57
22.	Chaita Pani Tea Garden	0.38
23.	Majua Tea Garden	0.24
24.	Upper Mamring Khasmahal	0.27
25.	Lower Mamring Khasmahal	0.32
26.	Toryak Khasmahal	0.58
27.	Mahal Diram Tea Garden	0.25
28.	Malotar Tea Garden	0.42
29.	Simring Tea Garden	0.36
30.	Sittong Khasmahal	0.98
31.	Barasit Tong Khasmahal	0.26
32.	Selpu Khasmahal	0.53
	Total	18.36

Table 32: List of Big Water Bodies in Siliguri Sub Division

Sl.No.	*Name of the Block*	*Name of the Water Body*	*Area (ha)*
1.	Phansideoa block	Ruhidasa beel	8.00
2.	Phansideoa block	Telengamoni beel	3.00
3.	Phansideoa block	Georamoni beel	1.00
4.	Phansideoa block	Sehira beel(Uttar)	0.09
5.	Phansideoa block	Sehira beel (South)	0.08
6.	Phansideoa block	Chak chaki beel	0.52
7.	Kharibari block	Harindoba beel	2.00
8.	Kharibari block	Kua-Khanda beel	1.50
9.	Naxalbari block	Rangapani beel	1.02
10.	Phansidewa block	Ghosh pukur beel	2.69
		Total	21.43

Seed resource survey and blockwise resource survey of Darjeeling district are presented in Table 33-41.

(*b*) Mirik Reservoir

Plate 10: Mirik Reservoir

The Mirik reservoir (latitude: 2653' N; longitude: 88 10' E; altitude: 5,296 ft) is now a tourist hotspot (Plate 10) in upland Darjeeling district. The Mirik reservoir (better known as Mirik or Sumendu Lake) is spread over 270 acres in the town it is named after and was constructed primarily to facilitate commercial tourism in Darjeeling district. The Mirik Tourist Project was commissioned on 7 March 1972 and 335.8 acres of swampland, earlier known as Sumendu Dhaap was requisitioned from Thurbo Tea Estate owned by the Goodricke group. The reservoir was inaugurated on 2 October 1979. The captivating beauty of the reservoir and its surroundings quickly transformed the tiny hamlet into a tourist hotspot. Many hotels and restaurants sprung up over the next decade along the banks of the reservoir (mainly Krishnanagar locality) and Mirik has since emerged as a prime tourist destination in Darjeeling district. However, the reservoir has suffered serious pollution problems associated with urbanisation and population increase. The maximum depth of the reservoir was 26 feet and minimum 6 feet when it was dug up. The situation is somewhat better now with some desiltation activities having taken place during 2003-04. Deforestation in the catchment areas and largescale construction activities along the banks has resulted in soil erosion. The study of hydrobiology of Mirik reservoir was estimated during 2000, and a census report from Mirik municipality, which indicated that the town had 15,162 nos. permanent residents. Nearly 50 per cent of the population resided very close to the reservoir. In

Plate 11: Salamanders Collected Adjacent to Mirik Reservoir

Table 33: Seed Resource Survey

Name of Stream	*Location*	*Distance Covered*	*Main Species*	*Availability of Seed*		*Average Size*	*Remarks*
				Peak	*Lean*		
Teesta	1. Kirney Khola 2. Bam Khola 3. Melli 4. Teesta quori 5. Teesta Bazar 6. 29th Mile 7. Sevoke	1 km 1 km 1 km 0.5 km 1 km 0.25 km 2 km	*N. hexagonolepis* *Schizothorax* sp. *Baralius* sp. *Danio* sp. *Tor tor* *Garra* sp.	Sept, (4185 pcs) Oct, (1561 pcs) Nov, (529 pcs) Dec, (378 pcs) Jan, (478 pcs) Feb, (484 pcs) Mar, (593 pcs)	April,(73 pcs) May, (Nil) June, (Nil)	5-15 cm	Teesta river stretch is mostly dominated by 5 main species. Harvesting was limited to cast net. *N. hexagonolepis* dominates the stock followed by *Schizothorax sp.* Teesta has various sp. in lower stretch around Sevoke area. Catch of *Angula bengalensis* was unique followed by coloured *Channa sp. Puntius sp.* was also present, the rivers adjoining the Teesta near Sevoke area. The catch was limited to cast net and use of hook and line found. Katli was caught by hook with maximum weight of 2.5 kg.
Reyang	1. Dumping area 2. Picnic area 3. Rambhi	1 km 0.5 km 1 km	*N. hexagonolepis* *Schizothorax* sp. *Baralius* sp. *Danio* sp. *Tor sp.*	Sept, (100 pcs) Oct, (100 pcs) Nov, (200 pcs) Dec, (5000 pcs) Jan, (6000 pcs) Feb, (7000 pcs) Mar,(15000pcs)	April,(50 pcs) May, (25 pcs) June, (Nil)	1-2 cm	Reyang is an important place for harvesting seed. Seeds of size 1-2 cm are in plenty. Most of the time from Nov, to March. This river is prone to illegal method of fishing like electro fishing and pesticides are rampant during post monsoon. Hence availability of large fish is not common. This river houses good nursery grounds, with clear water and gentle gradient. Mahseer and Katli were found in this river though in less no. but of good size (5-10 cm). Reyang has a good potential if developed as Coldwater Fish Sanctuary. There is no proper fishing gear so far developed for catching the fingerlings from their natural breeding ground.

Contd...

Table 33–Contd...

Name of Stream	*Location*	*Distance Covered*	*Main Species*	*Availability of Seed*		*Average Size*	*Remarks*
				Peak	*Lean*		
Relli	1. Picnic spot (upper) 2. Picnic spot (lower)	1 km 1 km	*N. hexagonolepis* *Schizothorax* sp. *Tor tor*	Dec, (200pcs) Jan, (500pcs) Feb, (500pcs) Mar, (2000pcs) April, (5000pcs)	May, (Nil) June, (Nil)	2-5 cm	Relli being far from our research station is a river where we have found max. no. of katli (*N. hexagonolepis*). Harvesting by cast net is not effective due to terrain nature of the river and fast current. Utensils were used for harvesting these fishes. It has been observed that the size of the fingerlings varies from place to place. It is also observed that the fish breeds wherever there is a favorable condition. Breeding starts from the middle of March as soon as the water temp starts rising during Sept. at various intervals. This was noted after the physical examination of the brood fish.
Rangeet	1. Beni 2. Jorthang road	2 km 1 km	*N. hexagonolepis* (Katli) *Schizothorax* sp. (Asala)	Dec, (100pcs) Jan, (1000pcs) Feb, (5800pcs) Mar, (8000pcs) April,(5000pcs)	June, July	1-5 cm	Rangeet which is clear, most of the time, is a place/area where harvesting of fish using cast net is unfruitful. Rangeet houses maximum number of *Schizothorax sp.* (Asala) and fewer no. of *N. hexagonolepis* (Katli). Since catching of fish by cast net was not effective, we settled to catch the fish using common/local technique in which utensils (local name MOKTU) are used. Though this is time and weather specific. We were successful in harvesting uncountable no. of Asala from this particular spot. The ideal day for the catch will be a sunny day, clear water and temparature ranging from 15 to 20 degree Celsius. Size of the fingerling ranges from 1.5 -2 cm weighing 5-50gms.

Table 34: Resource Survey of the Block–Jorebungow-Sukhiapokhri

Sl.No.	Name of the Village	No. of Fishermen Surveyed	Water Area (sqft)	Owner-ship	Water Depth (ft)	Species Culture	Duration of Culture (months)	Materials Input	Harves-ting Disposal in Kg. per Unit (i.e 1500 Sq. ft.)	Extension Support	Training	Labour Use	Credit
1.	Jue Bari	4	2900	Own	2–2.5	Common, Grass, Silver carp	6–9	Oil cake, Green leaves, Kitchen waste	180	Yes	Yes	Hired & Family	No
2.	Tarsinghdara	6	2300	–do–	2	–do–	5–10	–do–	192	–do–	–do–	Family	No
3.	Tipindara	7	3150	–do–	2–3	–do–	5–9	–do–	163	–do–	–do–	Hired & family	No
4.	Lamadara	3	900	–do–	2–2.5	–do–	6–9	–do–	159	–do–	–do–	Hired & family	No
5.	Chanong T.E.	1	750	–do–	2–3	–do–	5–10	–do–	189	–do–	–do–	Family	No
6.	Gairy Gaon	5	1700	–do–	2–2.5	–do–	6–9	–do–	165	–do–	–do–	Hired & family	No
7.	Dhajia	2	250	–do–	2–3	–do–	5–10	–do–	195	–do–	–do–	Hired & family	No
8.	Samrik	2	250	–do–	2–2.5	–do–	5–9	–do–	164	–do–	–do–	Hired & family	No
9.	Salembong	3	1050	–do–	2–3	–do–	6–9	–do–	187	–do–	–do–	Hired & family	No
10.	Soaek	2	350	–do–	2–2.5	–do–	5–10	–do–	154	–do–	–do–	Hired & family	No
11.	Plungdung	2	250	–do–	2–3	–do–	6–9	–do–	169	–do–	–do–	Hired & family	No
12.	Pokherebong	14	3050	–do–	2–3	–do–	5–10	Do, cow dung	153	–do–	–do–	Hired & family	No

Table 35: Resource Survey of the Block–Mirik

Sl.No.	Name of the Village	No. of Fishermen Surveyed	Water Area (sqft)	Ownership	Water Depth (ft)	Species Culture	Duration of Culture (months)	Materials Input	Harvesting Disposal in Kg. per Unit (i.e 1500 Sq. ft.)	Extension Support	Training	Labour Use	Credit
1.	Marma T.E.	11	2003	Own	2–2.5	Common, Grass, Silver carp	5–10	Atta, Oil cake, Green leaves, Kitchen waste	169	Yes	Yes	Hired & Family	No
2.	Thurboo T.E.	13	2325	Own	2–3	–do–	6–9	–do–	185	Yes	Yes	–do–	No
3.	Okayti T.E.	13	2478	Own	2–2.5	–do–	5–10	Atta, Oil cake, Green leaves, Kitchen waste	192	Yes	Yes	–do–	No
4.	Soureni Busty	8	1293	Own	2–3	–do–	6–10	–do–	180	Yes	Yes	–do–	No
5.	Mirik Busty	16	3274	Own	2–3	–do–	6–9	–do–	194	Yes	Yes	–do–	No

Table 36: Resource Survey of the Block–Kurseong

Sl.No.	Name of the Village	No. of Fishermen Surveyed	Water Area (sqft)	Owner-ship	Water Depth (ft)	Species Culture	Duration of Culture (months)	Materials Input	Harves-ting Disposal in Kg. per Unit (i.e 1500 Sq. ft.)	Extension Support	Training	Labour Use	Credit
1.	Mamring	5	867	Own	2–3	Common, Grass, Silver carp	6–9	Atta, Oil cake, Green leaves, Kitchen waste	185	Yes	Yes	Hired & Family	No
2.	Sivatar T. E.	10	2302	Own	2–2.5	Common, Grass, Silver carp	5–10	–do–	125	Yes	Yes	–do–	No
3.	Sittong	14	2420	Own	2–3	Common, Grass, Silver carp	6–10	–do–	175	Yes	Yes	–do–	No
4.	Gaurishankar T.E.	5	1376	Own	2–3	Common, Grass, Silver carp	5–9	–do–	192	Yes	Yes	Hired & Family	No
5.	Maldiram T.E.	4	770	Own	2–2.5	Common, Grass, Silver carp	6–10	–do–	160	Yes	Yes	Hired & Family	No
6.	Turuk Busty	14	2353	Own	2–3	Common, Grass, Silver carp	6–10	–do–	173	Yes	Yes	Hired & Family	No
7.	Singale T.E.	2	565	Own	2–2.5	Common, Grass, Silver carp	6–9	–do–	168	Yes	Yes	Hired & Family	No
8.	Kara Busty	1	192	Own	3	Common, Grass, Silver carp	9	–do–	165	Yes	Yes	Hired & Family	No

Table 37: Resource Survey of the Block-Kalimpong-I

Sl.No.	Name of the Village	No. of Fishermen Surveyed	Water Area (sqft)	Owner-ship	Water Depth (ft)	Species Culture	Duration of Culture (months)	Materials Input	Harves-ting Disposal in Kg. per Unit (i.e 1500 Sq. ft.)	Extension Support	Training	Labour Use	Credit
1.	Pudung	1	600	Own	2.5	Common, Grass, Silver carp	12	Oil cake, Green leaves, atta, Kitchen waste	144	Yes	Yes	Hired and family	No
2.	Sanook	1	1000	Own	3	–do–	9	–do–	166	–do–	–do–	–do–	No
3.	Pemling	17	5690	Own	3	–do–	6–10	–do–	188	–do–	–do–	–do–	No
4.	Sindebong	13	6440	Own	2–3	–do–	6–9	–do–	185	–do–	–do–	–do–	No
5.	Samalbong	4	1440	Own	2–3	–do–	6–12	–do–	150	–do–	No	–do–	Yes
6.	Soruk	6	3300	Own	2–3	–do–	6–12	–do–	165	–do–	Yes	–do–	Yes
7.	Samthar	2	700	Own	2–4	–do–	6–12	–do–	185	–do–	No	–do–	Yes
8.	Kankaybong	8	4150	Own	2–4	–do–	6–12	–do–	160	–do–	No	–do–	Yes
9.	Yok–Primtam	3	1950	Own	2–4	–do–	6–12	–do–	155	–do–	No	–do–	Yes
10.	Bong busty	7	2650	Own	2–4	–do–	6–12	–do–	195	–do–	No	–do–	Yes
11.	Balukhop	2	1000	Own	2–4	–do–	6–12	–do–	154	–do–	Yes	–do–	Yes
12.	Dungra	4	2250	Own	2–4	–do–	6–12	–do–	188	–do–	No	–do–	Yes

Table 38: Resource Survey of the Block-Kalimpong-II

Sl.No.	*Name of the Village*	*No. of Fishermen Surveyed*	*Water Area (sqft)*	*Owner-ship*	*Water Depth (ft)*	*Species Culture*	*Duration of Culture (months)*	*Materials Input*	*Harves-ting Disposal in Kg. per Unit (i.e 1500 Sq. ft.)*	*Extension Support*	*Training*	*Labour Use*	*Credit*
1.	Lolay	1	360	Own	3	Common, Grass, Silver carp	12	Oil cake, Green leaves, Kitchen waste, Rice bran	175	Yes	Yes	Hired & family	No
2.	Pabok	2	2400	Own	2–2.5	–do–	6–9	–do–	166	–do–	–do–	–do–	No
3.	Dhajay	5	1975	Own	2–4	–do–	5–12	–do–	193	–do–	–do–	–do–	Yes
4.	Payong	1	300	Own	2.5	–do–	6–9	–do–	145	–do–	–do–	–do–	No
5.	Kanjen	1	600	Own	3	–do–	10	–do–	157	–do–	–do–	–do–	No
6.	Dalapchand	6	5700	Own	2–4	–do–	6–12	–do–	190	–do–	–do–	–do–	Yes
7.	Gitdabling	4	1510	Own	2–4	–do–	6–12	–do–	183	No	–do–	–do–	Yes
8.	Kashone	2	1600	Own	2–4	–do–	2–4	–do–	170	Yes	–do–	–do–	–do–
9.	Vagay Pabak	7	5475	Own	2–4	–do–	6–12	–do–	160	Yes	Yes	–do–	Yes
10.	Monsong	3	1700	Own	2–4	–do–	6–12	–do–	170	Yes	Yes	–do–	Yes

Table 39: Resource Survey of the Block-Rangali Rongliot

Sl.No.	Name of the Village	No. of Fishermen Surveyed	Water Area (sqft)	Owner-ship	Water Depth (ft)	Species Culture	Duration of Culture (months)	Materials Input	Harves-ting Disposal in Kg. per Unit (i.e 1500 Sq. ft.)	Extension Support	Training	Labour Use	Credit
1.	Tinchulay	2	2000	Own	2–2.5	Common, Grass, Silver carp	6–12	Oil cake, Green leaves, Kitchen waste, Rice bran	165	No	No	Hired & family	No
2.	Hum busty	23	15300	Own	2–3	Common, Grass, Silver carp	6–11	–do–	180	No	No	Hired & family	No
3.	Ling ling	3	1800	Own	2–2.5	Common, Grass, Silver carp	6–12	–do–	189	No	No	Hired & family	No
4.	Rangoli bazaar	4	3000	Own	2–2.5	Common, Grass, Silver carp	6–12	–do–	187	No	No	Hired & family	No
5.	Takdah shree gaon	2	1000	Own	2–2.5	Common, Grass, Silver carp	5–11	–do–	197	No	No	Hired & family	No
6.	Takdah cant.	11	7900	Own	2–3	Common, Grass, Silver carp	6–12	–do–	155	No	No	Hired & family	No
7.	Lamahatta	18	13722	Own	2–3	Common, Grass, Silver carp	4–12	–do–	194	No	No	Hired & family	No

Table 40: Resource Survey of the Block-Darjeeling Pulbazar

Sl.No.	Name of the Village	No. of Fishermen Surveyed	Water Area (sqft)	Owner-ship	Water Depth (ft)	Species Culture	Duration of Culture (months)	Materials Input	Harves-ting Disposal in Kg. per Unit (i.e 1500 Sq. ft.)	Extension Support	Training	Labour Use	Credit
1.	Majua	3	625	Own	2–2.5	Common, Grass, Silver carp	4–9	Oil cake, Green leaves, Kitchen waste, Rice bran	166	Yes	Yes	Hired and family	No
2.	Kaijalay	10	2500	Own	2–3	Common, Grass, Silver carp	6–12	–do–	174	Yes	Yes	Hired & family	No
3.	Bijanbari	53	10210	Own	2–3	Common, Grass, Silver carp	4–12	–do–	161	Yes	Yes	Hired & family	No
4.	Chontong	4	1350	Own	2–3	Common, Grass, Silver carp	4–12	–do–	188	Yes	Yes	Hired & family	No

Table 41: Resource Survey of the Block-Gorubathan

Sl.No.	Name of the Village	No. of Fishermen Surveyed	Water Area (sqft)	Owner-ship	Water Depth (ft)	Species Culture	Duration of Culture (months)	Materials Input	Harves-ting Disposal in Kg. per Unit (i.e 1500 Sq. ft.)	Extension Support	Training	Labour Use	Credit
1.	Allay Busty	3	1350	Own	2–3	Polyculture	4–9	Atta, oil cake, Rice bran, kitchen waste, green leaves	170	Yes	Yes	Hired and family	yes
2.	Neem Busty	4	920	Own	2–3	Polyculture	5–10	–do–	154	Yes	Yes	Hired & family	Yes
3.	Gojitan	3	640	Own	2–2.5	Polyculture	4–10	–do–	150	Yes	Yes	Hired & family	Yes
4.	Ambiok	3	860	Own	2–3	Polyculture	4–9	–do–	160	Yes	Yes	Hired & family	Yes
5.	Jhandi Gaon	3	750	Own	2–2	Polyculture	4–10	–do–	158	Yes	Yes	Hired & family	Yes
6.	Lengsel	2	480	Own	2–3	Polyculture	4–9	–do–	140	Yes	Yes	Hired & family	Yes
7.	Anlay Busty	3	650	Own	2–3	Polyculture	5–10	–do–	135	Yes	Yes	Hired & family	Yes
8.	Sherpa Busty	9	2050	Own	2–3	Polyculture	4–9	–do–	160	Yes	Yes	Hired & family	Yes
9.	Godak	5	1170	Own	2–3	Polyculture	4–9	–do–	152	Yes	Yes	Hired & family	Yes
10.	Today tangta	1	150	Own	2–3	Polyculture	4–10	–do–	135	Yes	Yes	Hired & family	Yes
11.	Chichu	1	270	Own	2–2.5	Polyculture	5–10	–do–	140	Yes	Yes	Hired and family	Yes
12.	Karem Busty	1	255	Own	2–3	Polyculture	4–9	–do–	135	Yes	Yes	Hired & family	Yes
13.	Suruk	2	450	Own	2–3	Polyculture	5–10	–do–	139	Yes	Yes	Hired & family	Yes

other words, even if the tourist population is left unconsidered, the daily activities of about 7,000 people living so close to the reservoir should be influenced by their proximity to this water body in some way or the other. Moreover, their number could possibly have increased by now. Soil erosion has led to heavy siltation, which can be seen in the shape of delta-like formations at the mouth of open drains. In Mirik reservoir, there is no proper sewerage system in Mirik town and waste from the surrounding areas is discharged into the reservoir. The topography of Mirik is such that it is very difficult to plan a viable alternative for sewage disposal, since the reservoir is situated in a valley/depression and the populated areas are located either very near or just above the reservoir. Hence, gravitation would tend to pull anything towards the reservoir. This has led to a serious deterioration of reservoir water in certain pockets. Ecosystems are capable of assimilating specific amounts of various pollutants. After exceeding this capacity, considerable increases in respective concentrations occur and can cause eventual breakdown of the ecosystem structure. As sewage enters a reservoir, a series of problems related to the chemistry and biology of the reservoir begins. Entering from external sources, pathogenic organisms experience explosive growth and sanitary conditions rapidly deteriorate. The water undergoes intense changes and a decrease in quality. Because of high contamination by components released from the sediments, the water may no longer be feasible for drinking. In 2000, the nitrate-nitrogen values recorded from the Mirik reservoir always exceeded that limit. Another problem that has surfaced is the dumping of solid wastes, including plastics, from the surrounding localities, hotels and parks/lawns, which are always full of litter. Although desiltation activities in 2003-04 have changed the composition of the reservoir bed for the better, human habits seldom change and the tourists (along with local people) can be seen adding to the litter around even today. The water quality is greatly influenced by its intermixing with the reservoir bed, particularly the release of nutrients and other substances from bottom sediments into water and vice-versa a process that is hampered by the formation of a plastic layer between the water and bottom soil.

(c) Management Guidelines for Mirik Reservoir

Reservoirs are man-made lakes and deserve their name because they are artificial lakes created by man for particular purposes. This creation makes reservoirs different in many aspects from lakes; thus, several aspects of their management are different from general lakes and the reason for the reservoirs construction should be taken into consideration while planning management strategies. Listed below are some management suggestions that could be considered by policy makers/authorities concerned with the Mirik reservoir:

(*i*) The capacity of any reservoir to accommodate recreation is not unlimited and certain restrictive rules must be put in place.

(*ii*) Strict policing along the reservoir shoreline and severe penalties/fines imposed on anyone caught littering (or not using dustbins) would curb the solid waste dumping problem to a large extent;

(*iii*) A proper sewerage system should be developed for Mirik town. Planners must ensure that wastewater does not enter the reservoir. If, in some cases, there is no alternative but to open up some drains into the reservoir, certain ecotechnological tools could be applied to clear up the wastewater first.

(*iv*) A proper system of sluice gates could be arranged from where the wastewater first enters small wetlands/tanks where aquatic weeds are cultured. These tanks could act as buffer zones where the excess nutrients from the wastewater would be utilised by the vegetation for their tissue-growth and water quality is restored to some extent before entering the reservoir;

(*v*) There are examples from different regions around the world of biomanipulation in the control of water pollution problems. Presently, only a few varieties of fish (common carp and grass carp) exist in the Mirik reservoir. Fish diversity could be increased or manipulated to cover the different habitats within the reservoir.

(*vi*) Problems of water quality deterioration start when there is an increased availability of nutrients for algae growth.

(*vii*) Eutrophic reservoir produces much more plant material that can be used by herbivores. This surplus accumulates in the lake until it is decomposed by bacteria, leading to a progressive decline of dissolved oxygen in the deeper strata, which varies considerably with an increase in pH and dissolved oxygen in the surface water, following the consumption of carbon dioxide by photosynthesis. These variations in the chemical environment severely damage the natural ecological rhythm of the reservoir.

(*viii*) Recreational fishing as a means of tourism, has not received proper attention in India, however, if properly organized, it can only be a matter of time before hobbyist anglers come to Mirik to enjoy angling.

(*ix*) The cost of reservoir restoration can be at least partially generated from the reservoir itself, if local tourism managers create a proper environment for boating, angling, etc. Boating facilities already exist in Mirik, however needs proper management and introduction of advance fibre boats so that every tourist coming to Mirik should feel the urge to sail in a boat for some time.

(*x*) The formation of sport fisheries association could be encouraged.

(*xi*) Although only a hypothesis at present, it could be argued that if a variety of herbivorous fish could be collected from local jhora/hill streams and stocked in the reservoir, the problem of excess algae growth could be curtailed.

(*xii*) Besides, with proper planning and management, the reservoir can serve as an excellent rearing venue for coldwater sport fish species;

(*xiii*) Fishing is considered one of the most popular forms of outdoor recreation throughout the world. Reasons for its popularity vis-à-vis other outdoor activities are the relatively low cost and lack of physical demands.

(*xiv*) Fish can be collected from local rivers like the Balasun, Mahananda and Teesta, stocked, and reared in the reservoir.

(*xv*) Recreational anglers can be given licenses for angling in the reservoir at a reasonable fee for a specific period.

(*xvi*) The money generated from anglers would be useful in maintaining the fish stock and in carrying out renovation activities of the reservoir from time to time.

People who live around a lake/reservoir generally feel some kind of affinity with the lake/reservoir environment. There has been evidence of local people helping in environment restoration activities around certain lakes. In the Great Lakes Basin (North America), active citizens played a major role in setting the agenda for cleaning up the Great Lakes. At an International Joint Commission meeting in 1989, Dr Mark van Putten of the National Wildlife Federation in the USA summarized the results of the past decade of citizen activism in respect of people involvement in the development and maintenance of lakes. Mirik reservoir is only about 25 years old and the population surrounding it has mostly settled in over the past two decades. The local community should be educated about management procedures. With more and more anglers and tourists coming to this serene destination, the local people being the beneficiaries would get motivated and encouraged to participate in the development programmes. Only the local community can ensure a check on illegal poaching, littering and other deteriorating influences in the lake.

To the extent that fishing tourism is successful and more anglers travel for this recreation to Mirik in future, it will be increasingly important to consider the social, economical and environmental impacts of this industry. The purpose of nature tourism is to watch with joy scenes of nature, including wildlife. In an article published in the highly acclaimed journal, Biodiversity and Conservation, Dr H Goodwin has defined ecotourism as low impact nature tourism that contributes to species or habitat maintenance either directly through a contribution to conservation and/or indirectly by providing revenue to the local community, sufficient for local people to value and therefore protect their wildlife heritage (reservoir habitat in this case) as a source of income. Because of its consumptive nature, many scholars would not consider fishing as ecotourism. However, a case has been made by Dr SM Holland and his colleagues from the Centre for Tourism Research and Development, University of Florida, USA, that some forms of fishing can be viewed as ecotourism. In these cases, fishing is practiced on a catch and release basis, and provides positive economic benefits (employment, incorporation of new revenues) for local communities and direct financial support (fishing license fees, entrance fees, etc.) to the local area. Even more importantly, it seeks to foster ethical behavior that protects and sustains the fish population, provides political support for clean water status and local conservation initiatives and promotes a respect for nature. The involvement of the local community (hill people) in developmental initiatives can also bridge the feeling of alienation.

(*ii*) Jalpaiguri District

This district has got its name from the famous 'God Jalpeswar' or its richness in the fruit 'Jalpai' or Olive. It consists of Ucchabach forest, Tea garden and Duars (Torai). Southern part of this district is plain. Mahananda on the west and Sonkosh River is in the east refers the boundary of the district. Except hilly forest areas on the east, majority of the soil of this district is clay-loam. Jalpaiguri is full of rivers. The geographical area of Jalpaiguri district is 6227 square kms. out of which 28.75 per cent area, *i.e.* 1790 sq km. is recorded as forest land. It has a rich diversity of flora and fauna having around 700 general and 2000 species of dicotyledonous plants. Collectively all the water bodies form important sources for cultivation of fishes and development of fishery industry in this zone. A number of tribes dwell among the natural excavations of this district and have less means of livelihood support – may utilize this water source to form an objective of livelihood and may be trained to become skilled fisherman thus leading to financial and industrial upliftment of the area. Being situated on the foothills of the great Himalayas it has a number of rivers flowing from North to South. In addition to a fair number of rivers (Figure 5) flowing in the district, namely Teesta, Torsa, Jaldhaka, Kalyani, Sankosh, Raidak, Dina, Dudua, Dana, Dhorla, Korla, Jamuna this district also bears a number of lakes and water reservoirs, beels etc. many of which are of historical importance.

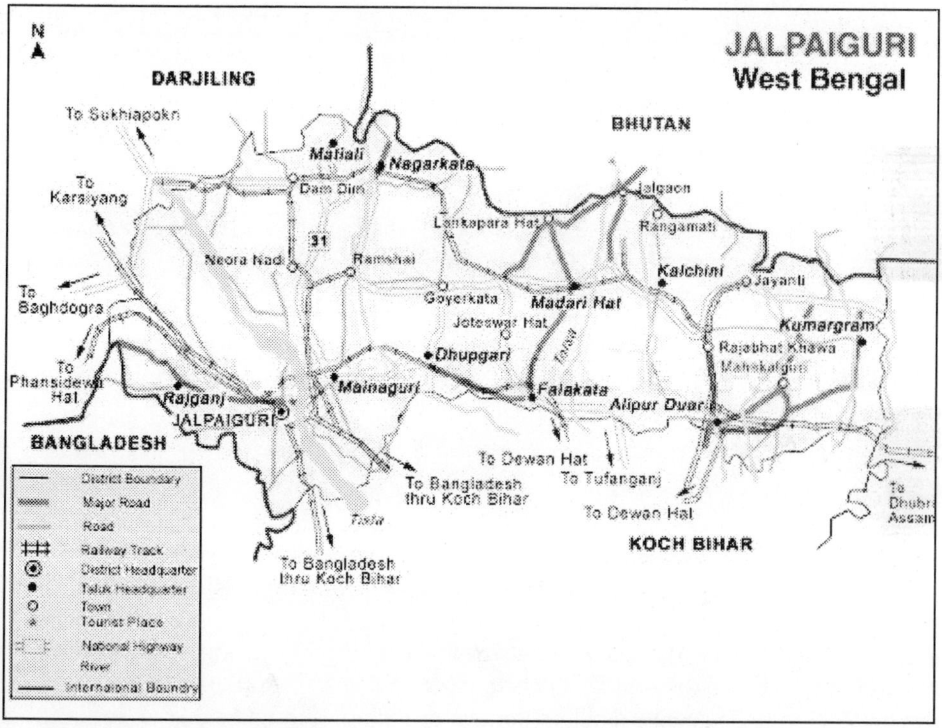

Source: www.mapsofindia.com

Figure 5: River Map of Jalpaiguri District

a. Rivers and Big Waterbodies

Teesta

In the map of Renel (1800 century) Teesta was shown to make the path of Kortoa river. It seems that Teesta falls in Padma along with Atrai in Dinajpur by this path. In the flood of 1968 everything went hay wired. Leaving the ancient path, Teesta tries to flow through a small river. At last Teesta traced its new path itself and falls in Brahmaputra. This new path was little bent. Teesta's ancient path is known as Choto

Teesta, Mara Teesta, Buri Teesta etc. Like Ganga and Brahmaputra, Teesta carries huge amount of pollution. The district has a lot of tributary of Teesta. Some of them are:

Lish

Leaving Dargeeling district falls in Teesta near Bagra kot.

Gish (Git khola)

Passing forward by Odlabari tea garden, falls in Teesta near Hanskhali. The length of this river in the district is nearly 21 km.

Chel

Flowing on the east of Choto rail line in Chel and on the west is Gish. This rivers too, leaving Odlabari tea garden amidst forest areas.

Karala (Kalla)

It originates from somewhere in the forest of Baikunthapur and meets Teesta near Jalpaiguri city. This is a right tributary of Teesta.

Dharala

It originates from the southern area of lower Tandu reserve forest. Flowing down towards south, passing by Neora rail station it gets united with Neora river. Latter it falls in Teesta being united with Chel.

Neora Khola

Unknown hilly river Neora has become famous at present for its water project. Neora is the river of Kalimpong hills. Passing through deep forest Neora khola flows down to Garubathan. This forest has great importance for its rare species of flora and fauna. Up to Garubathan the length of the river is 58 km.

Kartoa

In puran this river has been described as sacred. In Mahasthangarh broken statue of Goddess Kartoa has found. Bathing in river Kartoa on occasion is believed to be very sacred. Ancient to unit Hu-en-sung left for Pragjotishpur crossing this river. But it is difficult to say which flow he crossed. In ancient times, Kartoa has known as Sadaneera. United Kartoa–Teesta was the main water path in ancient North Bengal. Kartoa is now a separate river. Arising, from Baikunthapur forest, flowing to Bangladesh, now it falls in Jamuna as Fuljhur. But Teesta has no connection with Kartoa.

Jaldhaka

One of the largest river of Jalpaiguri district is Jaldhaka. Arising from Bitang lake area of Sikkim (height 4420 mt.) Dichu and Lichu river gave birth to Jaldhaka being united with some other rivers. The river flows to Coochbihar passing by Nagrakata, Ramsai and Amguri. Jaldhaka's main right tributary in this district is Murti. Murti is nearly 24 km long.

Dayna

Jaldhaka's main left tributary is Dayna. Its bornplace is Bhutan hills. Dayna falls in Jaldhaka leaving Ramsai town. Dayna is very flood-prone.

Dudhuna

It is also a big river. Angar basa, Nanai, Goir kata and some other small rivers give birth to Dudhuna. All arise from the forest of Duarse. Dudhuna falls in Jaldhaka flowing to Coochbihar.

Mujnai

Arising from southern place of Bhutan hill, leaving through Totopara and flowing down Jalpaiguri district it goes to Coochbihar. One tributary of this river is called Pagli.

Torsha

Like Teesta and Jaldhaka one famous river of Jalpaiguri district is Torsha. Its name in Chumbi and Bhutan is Amo – Chu. Torash – Toi Rosha means 'angry waterfalls'. Leaving Funt, shoaling, flowing nearby Joygaou enter Jalpaiguri district. Later passes to Coochbihar by Hansimara, Falakata way. It is a dangerous river. It has diversed into two branches near Hansimara. East one is shilly and the west one is Char Torsha. In Coochbihar they are united into one. It is difficult to say which one will become violent in incessant rain. They are 40 km long in this district. Main tributaries of Torsha in this district are – Malengi bela, Sunjai, Choto torsha (Halang), and Kala torsha.

Kaljani

One big speedy river named Alaikuri arising from Bhutan hill, flowing down towards south meets Deema river near Alipurduar. Deema's birthplace is Baksha hill. Some important tributary of Kaljani are – Gabar basra, Buri basra, and Bania. Nimti jhora, and Parore are important tributaries.

Raidak

On the north of Bhutan, Tibbet's Chomalhari mountain is the birthplace of this river. Its name in Bhutan is Chim – Chu. Coming through Bhutan ghat, it makes its entry to Jalpaiguri district.

Sonkosh

Arising from Panakha (or Punakha) hills of Bhutan flowing down as Punkha chu it enters Jalpaiguri and takes new name Sonkosh. Western bank of Sankosh is taken as border of West Bengal and Assam. Besides above mentioned rivers, there are some other less known small rivers in Jalpaiguri district. Some of them are Chitijhora, Panga, Kalapani, Birkati etc. In spite of being small in size they become dangerous in monsoon period and often overflow to nearby areas causing huge flood.

The block-wise list of big water bodies of Jalpaiguri district is depicted in Table 42.

There are 74 tribal villages in Jalpaiguri district. Subsidy is mainly on agriculture and employment in forestry including timber operation, plantation works etc. Some of the villagers maintain cow and poultry. Many villages hence become members of Large Singled Agricultural Multipurpose Cooperative Societies (LAMPS).

As there are a large number of water bodies in this district hence fishing industry and manufacture of fish products have a good scope. This provides opportunities, for investment in this sector as well. Low lying areas and river sides have been successfully utilized by local people for cultivation of fish in this district. Organized effects in this direction would be fruitful. Fish pond ecodyke system for integrated fishery and sericulture projects also has potentiality in this area.

Table 42: Block-wise List of Big Waterbodies of Jalpaiguri District

Sl No.	Name of the Block	Name of the Water Body	Area (ha)
1.	Dhupguri MRL-2.6	Gilandi lake	FRL–10.5
2.	Dhupguri MRL-6.75	Jorapani beel	FRL–13.5
3.	Kumargram	Kalapani Beel MRL-8.0	FRL–12.0
4.	Alipurduar-I MRL-15.91	Ratneswar Jheel	FRL–20.0

FRL: Full Reservoir Level; MRL: Minimum Reservoir Level.

b. Utilisation of Big Waterbodies (Dighis) for Benefit of the Scheduled Caste, Tribe and Other Backward Class

(a) Pisciculture in the Relevant Area

The district covers an area of 62.45 sq kms. among which cultivated lands occupies 55, 820.4 ha. The huge numbers of beels are of extreme importance as a source of natural property. Many ponds due to siltation through 100 years have become unfit for fish cultivation. Due to this agricultural crop is being planted here instead of fish. At this point it should be noted that in these areas water bodies have not been dug out solely for the purpose of Pisciculture. Since ancient time a fair variety of fishes were available in this region from natural water sources. In addition among the Rajbangshi sect, selling of fishes was a criticisable offence. With the advent of modern era demand for edible fishes has increased and Pisiculture is being practiced in a number of tanks like 'Amgury's Dangi Dighi' Mahit Nagar's Jordighi, Rajbari's Dighi, Kanchan Dighi etc. Thus it is being proposed that many more such big water bodies of this district can be renovated, reclaimed and finally established for fish cultivation purposes, through formation of Fisheries co-operative societies which in turn will:

1. Cause fellow ponds and wetlands to be converted into useful beels and 'wealth' lands.
2. Cause employment generation through training young generation and traditional fishers.
3. Solution of the ownership problem of the water bodies.

Goralbari, Bahadur, Sukhani, Mondalghat areas are undeveloped and lie at the border area of the district and bear immense scenic beauty. In Johuri Talma area many archeological remnants of Pal and Sen Era have been discovered. Moinaguri block excel in temples and other historical archives. 99th capital of prince Govindah, son of queen Moinamoti, remnants of fort of king of Chilpata, Boxafort, Makrapara kali temples also embrace this district. Development of 'Dighis' residing will help in financial as well as cultural upliftment of the district due to formation of important historical tourist spot.

(b) Lotus Cultivation

Natural lotus flowers bloom in many ponds *e.g.* in Moinaguri, Aamgury, Sadar, Rajganj. Interference of Agriculture and Horticulture Depts. of the state required to cultivate lotus in commercial scale.

(c) Social Forestry-cum-Fishery

In addition to fish, duckery, ranicrop, social forestry may also be encouraged.

(d) Culture of Medicinal Plants

Long extensions (approx 100 bighas) of fallow lands are lying in the embankment of the lake since centuries. This will form an ideal area for cultivation of medical plants–leading to establishment of medical industry in this area. Availability of water to cultivate fish and other crops will not be a major problem. Study of livelihood, pattern, socio-economic status, religion, habit, culture need also to be studied and developed.

(e) Fishery System Already Existing in Jalpaiguri District

1. Culture of integrated fishery with Khaki Cambel duckery.
2. Manufacture of fish food from poultry excretory products.
3. Using aerator for supply of sufficient quantity of oxygen to the cultivated fish.
4. Culture of *Clarius batrachus.*

(f) Future Prospect

1. Multidimensional fishery industry.
2. Jhora fish culture in hilly region.
3. Fish culture in waste water fed channels of the town.
4. Cage culture of fish.
5. Pen culture.
6. Culture of *Tilapia mosambica.*

Culture of fish in wetlands of undecided ownership is temporarily ideal occupation for weaker section of the people.

(*iii*) Coochbihar District

Most probably the name came from the population "Coach". The district Coochbihar is 3386 sq. km. The forest with hills covers district. Dominant part of the northern area is 'Torai'. In the east and west side of Coochbihar there are two rivers

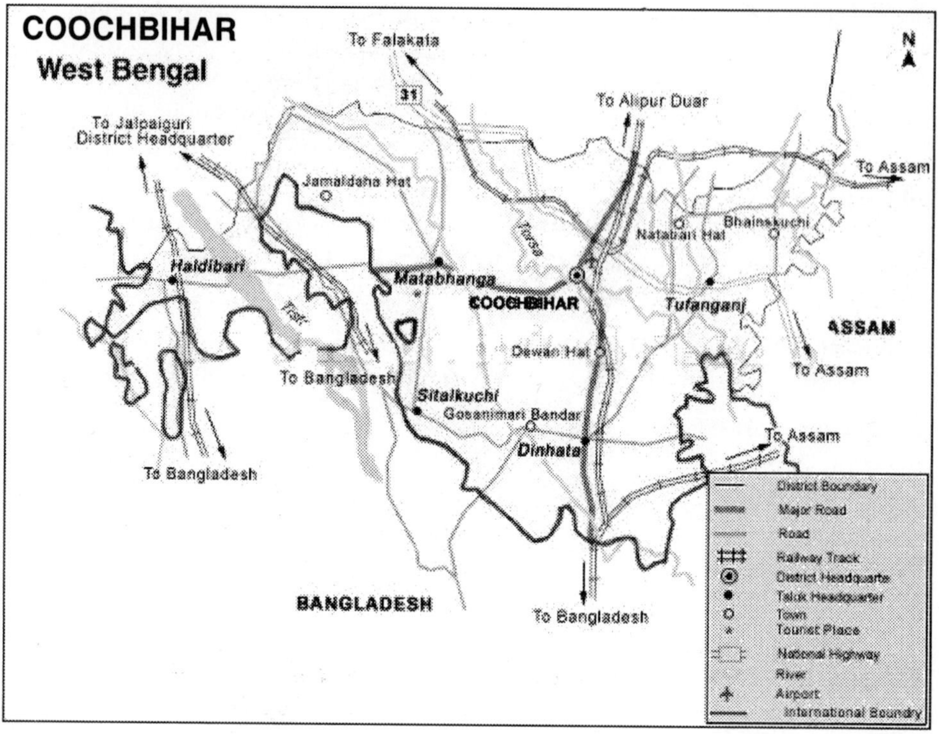

Source: www.mapsofindia.com

Figure 6: River Map of Coochbihar District

namely Sankosh and Teesta respectively. The rivers (Figure 6) of this district are briefed in below.

Teesta

The river Teesta enters in Coochbihar at Bakshiganj, then mixed with the river Brahmaputra and cross over the Kharkharia, Jharsinheswar, Mehkligaj, Kuchlibari etc. The total length in W.B. is 269 km. and it is only 13 km in Coochbihar. This is a perennial river.

Bura (or Buri) Teesta

This is a stream from the west side of Teesta, filled with rain.

Sutunga

This is a stream of river Jaldhaka; enter in this district at Kamatchangara Bandha and in the town Mathavanga it mixes with Teesta.

Jaldhaka

Many names of this river are Singmari, Mansai, Mujnai, Dankhan etc. in Coochbihar, mixed with river Torsha. Tributaries are – Dharla, Gidhari, Dolang and Dudhuya. Jaldhaka is known as Singari in Coochbihar.

Gidhari

Comes from Bangladesh and flow over Sitalkuchi, Lalbazar and mixed with torsha. This is the right tributary of Jaldhaka.

Dudhuya

Crossing over Coochbihar district, it mixed with Jaldhaka at Phulbari.

Dolang

Starts from the border of Jalpaiguri and Coochbihar. Left tributary of Jaldhaka, most probably is the old flow of Teesta.

Mujnai

Left tributary of Jaldhaka, mixed with Jaldhka at Rangamati in Coachbihar.

Torsha

Named as Dhala or Bara torsha. Enter into Coochbihar at Lafabari. Town Coochbihar is situated at the bank of this Bura torsha. In east–south of this district it mixed with Kaljani river. Then the mixed flow added with river Brahmaputra in Bangladesh at Moglahat. In Jalpaiguri two tributaries of Bura torsha are Mara torsha and Ghargharia.

Kaljani

Kept Alipurduar at left, enter in Coochbihar at Kholtar. Left Coochbihar at Jhaukuthi, mixed with Sankosh and then Brahmaputra in Bangladesh. Two tributaries are there – Gadadhar (or Choto Gadhadhar) and 1 No. Raidak (Bara Raidak). Gadadhar enter in this district at Khasbas and mixed with Kaljani at Siliguri. Bara Raidak is also mixed with Kaljani at Siliguri. Then the river is known in so many names as – Kaljani, Raidak, Dudhkumar, Sankosh etc.

1 No. Raidak: (or Bara Raidak)

Enter in this district at Ghentimari and in the near Siliguri open into river Kaljani and ultimately mixed with river Raidak. Popular tributary is Deepa (or Dipu).

2 No. Raidak

Enter in this district at Rampurhat, mixed with Sankosh at Sahebganje.

Sankosh

Flowing from the east side of Coochbihar, mixed with 2 No. Raidak or Choto Raidak and then named as Bara Gadadhar (or Gangadhar).

Except the above mentioned rivers, there is few other rivers *viz.* Gilandi, Chenakata, Sannasikata, Jerai, Saliajan, Chotomansai etc in Coochbihar.

The block-wise list of big water bodies of Coochbihar district is depicted in Table 43.

Table 43: Block-wise List of Big Water Bodies (Beel) in Coochbihar District

Sl.No.	Name of the Block	Name of the Beel/Baor	Total Area (Acre)
1.	Coochbihar-I	Dalipath Beel	19.35
2.	–do–	Houserdera Nadi	13.77
3.	–do–	Baiganbari Chhara	11.75
4.	–do–	Mansai Nadir Chhara No.-4	12.70
5.	–do–	Bharat Nadi with Masankura and Bairati Nadir Chhara No. 6	40.36
6.	–do–	Kanchi Nadi No. -13	13.24
7.	–do–	Sanyasikura (Hagura Nadi) No. 23	89.89
8.	–do–	Dharla Nadir Chhara No. 27	34.04
9.	–do–	Dharla Nadir Chhara No. 28	23.00
10.	–do–	Dharla Nadir Chhara No. 29	129.91
11.	–do–	Mara Mansai Nadi No. 34	71.72
12.	–do–	Kushari Nadir chhara No. 16	35.21
13.	–do–	Mara Mansai Nadi No. 40	51.60
14.	–do–	Rangasing Nadi No. 39	17.58
15.	–do–	Mara Mansai Nadi No. 42	72.40
16.	–do–	Ghargharia Nadi No. 75	38.46
17.	Coochbihar-II	Kaljani Nadir Charraot II St. No. -1	12.27
18.	Tufanganj–I	Maradanmga and Jadukhera River St. No. -1	19.21
19.	–do–	Chakuli, dahar, Patbanga& Ichai Nadi St.No. 2	134.74
20.	–do–	Kaljani Nadir Chhera Jhapoi Nadi and Chekadara Nadi St. No. 4 and 6	159.30
21.	–do–	Kaljani Nadir Chhera St. No. -9	52.37
22.	–do–	Tibet Nadir Chhera St. No. 6	33.1
23.	–do–	Bannaguri Nadi St. No. 30	26.00
24.	–do–	Raidak Nadir Chhera St. No. 36	32.06
25.	–do–	Raidak Nadir ChheraSt. No. 49	136.26
26.	–do–	Ayarani Chitalia beel and Salmara Nadi St. No. 1(d)	18.25
27.	Tufanganj–II	Bamni Nadi St. No. 37	16.84
28.	–do–	Raidak Nadir Chhera St. No. 13	26.78
29.	–do–	Beser Chhara St. No. 17	35.05
30.	–do–	Chand Gaon Nadir beel St. No. 39	43.87
31.	Dinhata–I	Singhimari Nadir Chhara St. No. – 1	38.34
32.	–do–	Singhimari Nadir Chhara St. No. – 4	135.00
33.	–do–	Darla Nadir Chhara St. No. 7	110.67
34.	–do–	Bura Dharla Nadi Chhara st. No. 9	490.31

Contd...

Table 43–Contd...

Sl.No.	*Name of the Block*	*Name of the Beel/Baor*	*Total Area (Acre)*
35.	Dinhata–I	Kaldhoba Chhara Nadi Balakura St. No. 15	13.01
36.	–do–	Jari Bura Dharla Nadi St. No. 69	32.33
37.	–do–	Jari Dharla Nadi Part II St. No. 75	113.40
38.	–do–	Dharla Nadir Chhara and Kura St. No. 32	42.17
39.	–do–	Jari Baniadaha Nadi Part I St No. 70	34.11
40.	–do–	Jari Baniadaha Nadi Part II St No. 71	41.07
41.	–do–	Jari Baniadaha Nadi Part II St No. 72	41.09
42.	–do–	Jai Mansai Nadi Part III St. No. 73	31.77
43.	Dinhata–II	Langulia Nadi and Puskarini St. No. 27	46.15
44.	–do–	Kalir Chhara and ParanerChhara St. No.30	37.12
45.	–do–	Karola Nadi, Haldikura Nadi, Jhalagalia Nadi and Fulkumar Nadi St. No. 24	58.02
46.	–do–	Abhichari Nadi and Jala St. No. 33	32.02
47.	–do–	Khursabeel and Bnara Bangdakir Beel St. No. 38	51.13
48.	–do–	Chaiter Chhara Beel and Jhowkuhi Beel	86.08
49.	–do–	Kanyamati Beel, Khotamara Nadi, Khektikura Beel St. No. 52	25.05
50.	–do–	Kachuani Beel and Khutamara beel St. No. 53	89.21
51.	–do–	Mansai Nadir Garai Chhara Nadi St. No. 61	41.04
52.	–do–	Sahebganj Nadi and Jala St. No. 65	42.04
53.	–do–	Telanir Deba, Fulkumar Nadi and Gharamara St. No. 68	30.14
54.	–do–	Krishnakura Kalaighati Beel St. No. 66	18.07
55.	Mathabhanga – I	Dhaulkurdanga Nadi st.No. 13	14.43
56.	–do–	Marasutunga Nadi St. no. 11	22.62
57.	–do–	Marasutunga Nadi St. no. 47	16.16
58.	–do–	Madaijhar nadi st. No. 9	17.16
59.	–do–	Marasutunga Nadi St. no. 12	26.90
60.	–do–	Sewli Nadi St. No. 17	42.63
61.	–do–	Kachua Nadi St. No. 8	17.95
62.	–do–	Sutunga Nadi St. No. 47/a	16.16
63.	Mathabhanga – II	Gelandi Nadi St. No. 59	26.00
64.	–do–	Jari Jaldhowa Nadi St. No. 49	23.00
65.	–do–	Prathamar Chhara St. No. 27	19.00
66.	–do–	Mansai Nadir Jal Chalachaler Rasta St. No. 28	15.50
67.	–do–	Jari Mansai Nadi St. No.Part II, III, St. No. 50,51	27.00
68.	Sitai	Maldanad St. no. 2(I)	88.51

Contd...

Table 43–Contd...

Sl.No.	Name of the Block	Name of the Beel/Baor	Total Area (Acre)
69.	Mathabhanga – II	Jai Ratnai Nadi St. No. 7(I)	19.75
70.	–do–	Gidari Nadi St. No. 6(I)	89.96
71.	–do–	Sitai Beel	37.57
72.	–do–	Jari Mansai Nadi St. No. 1(d)	76.47
73.	–do–	Singimari Nadi St. No. 8(I)	137.28
74.	Sitalkuchi	Khutamara Nadi St. No. 10	35.55
75.	–do–	Singimari Nadler Chhara St. No. 1	6.91
76.	–do–	Singimari Nadi St. No. 12	10.30
77.	–do–	Khutamara Nadi St. No. 12	10.94
78.	–do–	Khutamara Nadi St. No. 4	20.00
79.	–do–	Bura Dharla Nadi St. No. 5	5.36
80.	–do–	Mansai Nadi Chhara St. No. 14	15.90
81.	–do–	Dhar Mansai Nadi St. No. 2	51.27
82.	–do–	Kachua Nadi St. No. 6	30.84
83.	–do–	Jari Dharala Nadi Part II St. No. 18	33.06
84.	–do–	Jari Dharla Nadi Part I St. No. 17	5.86
85.	Haldibari	Girai Nadi St No. 7	30.63
86.	–do–	Buri Tista Nadi St. No. 6	11.01
87.	Mekhliganj	Jari Soniajan Nadi St. No. 21	68.38
88.	–do–	Jari Dharla Nadi st. No. 28	23.43
89.	–do–	Jala sua Nadi St. No. 27	60.71
90.	–do–	Nawanikura Nadi St. No. 13	22.66
91.	–do–	Sutunga Nadi st. No. 26	64.54
92.	–do–	Seuli Nadi St. No.	19.74
		Total Area (Acre):	4360.56

(*iv*) Dinajpur (North and South) District

Total area of Dinajpur is 5206 sq. km. From 1992 it is divided in to two districts, one is north Dinajpur and other is south Dinajpur. The clay-loam soil of Ganga and others rivers (Figures 7 and 8) formed hard-clay soil layer, known as 'Barind'. The upper portion is yellowish and lower is brackish red. Geographically it is divided into two regions–one is 'Duarsh area' and other is 'Dowab'.

Mahananda

Flowing through the district Purnia or Bihar towards the border of Dinajpur, it mixed with river Nagar Ituhar. Then the mixed stream named as Mahananda, flowing 32 kms upto the border of Malda and Dinajpur.

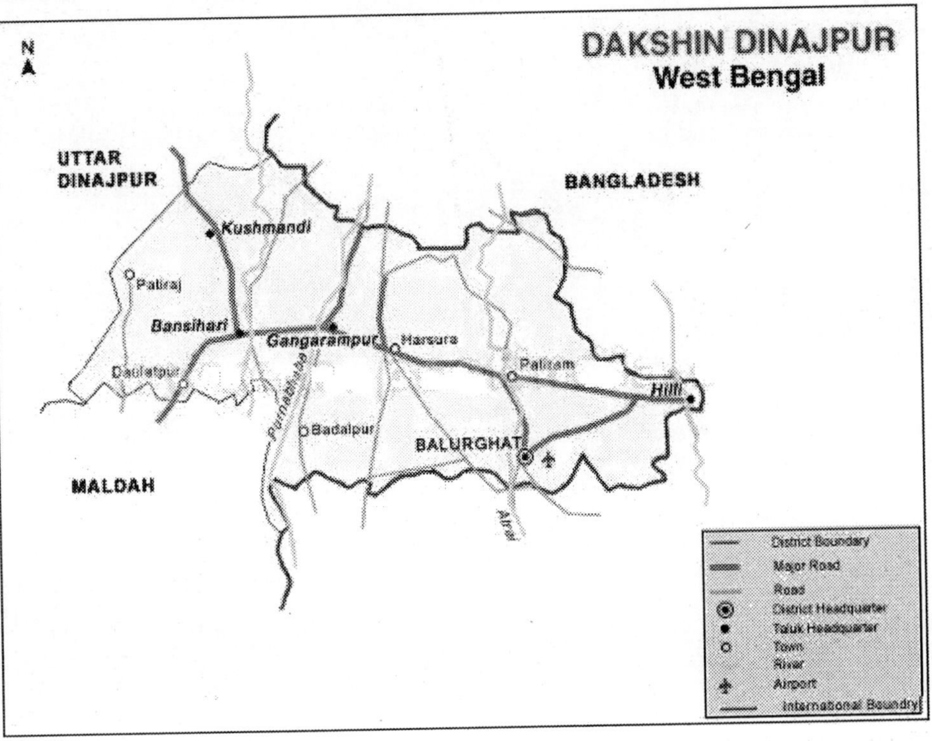

Source: www.mapsofindia.com

Figure 7: River Map of Dakshin Dinajpur District

Nagar

Arise from the lower aquatic region of Dinajpur, Purnia and Jalpaiguri, mixed with Mahananda, length 144 km.

Gamari (or Gamar)

Flowing from a 'Beel' near Kaliaganj.

Kulik

There are so many tributaries are of river Nagar, but Kulik is most popular. In this district the length of Kulik is only 11 kms.

Chirmati (Cheramati or Shrimati)

Flowing on Kaliaganj, Itahar, Bangsihari towards Malda. The length of the river in this district is 48 kms.

Tangan

Enter in this district at Kaliaganj and left at Gangarampur towards Malda. It is also a perennial tributary of Mahananda. This river was connected with the river Punnarvaba by a channel.

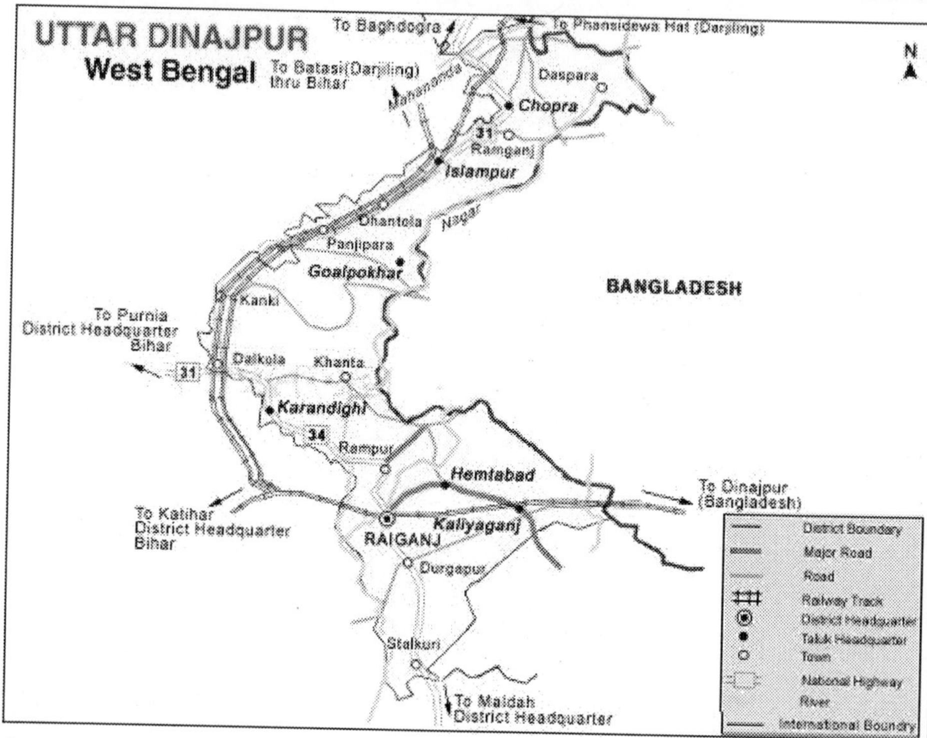

Source: www.mapsofindia.com

Figure 8: River Map of Uttar Dinajpur District

Punnarvaba

This is one of the best rivers of Dinajpur. This river is not too much long. Comes from Bangladesh enter in this district at Gangarampur then Malda to Bangladesh. Brahmrani is also a tributary of Punnarvaba.

Atrai (Atreyi)

Enter in Dinajpur at Chandpur and then ultimately open with river Padma at Bangladesh. Most probably Atreyi was the main steam of river Teesta. But from 1968, Atrayi was separated from Teesta. Only one tributary of Atrayi named Ichamati flowing in this district.

Except this rivers, there are so many small rivers are in Dinajpur. They are – Dahuk, Khari, Pitanu, Sudhani etc.

The block-wise list of big water bodies of Dakshin Dinajpur and Uttar Dinajpur districts are depicted in Table 44 and 45.

Table 44: Block-wise List of Big Water Bodies of Dakshin Dinajpur District

Name of the Block	*Name of the Water Body*	*Status*	*Area in Ha*	*Remark*
Balurghat	Ramproad Dighi	Private	3.20	Culturable
	Malancha Dighi	Private	2.80	Culturable
	Kalidaha Abra Beel	Vested	8.00	Derelict
Hili	Matia Pukur	Private	5.78	Semi Culturable
	Taldighi	Private	6.45	Culturable
	Gachina Beel	Partial Vested	12.00	Semi Culturable
	Bhaluka Beel	Partial Vested	11.70	Derelict
Kurnarganj	Kajala Beel	Vested	9.77	Semi Culturable
Tapan	Tapan Dighi	Vested	32.00	Semi Culturable
	Kadma Dighi	Vested	12.00	Culturable
	Rampur Dighi	Private	4.00	Culturable
Gangarumpur	Pathari Beel	Vested	25.00	Derelict
	Pran Sagar Dighi	Vested	18.23	Culturable
	Sapran Dighi	Partly Vested	8.00	Derelict
	Dhal Dighi	Partly Vested	26.00	Culturable
	Kaldighi	Zilla Parisad Owned	32.00	Culturable
	Fokar Beel	Vested	6.38	Derelict
	Sonar Kura Beel	Vested	8.80	Derelict
	Chikidaha Beel	Vested	11.65	Derelict
	Sutki Beel	Vested	4.80	Derelict
Bangshinari	Nandan Beel	Vested	8.40	Derelict
	Karal Beel	Vested	8.05	Derelict
	Baram Beel	Vested	4.67	Derelict
Harirampur	Sankihar Dighi	Vested	4.54	Culturable
	Gour Dighi	Vested	27.24	Semi Culturable
	Malian Dighi	Vested	32.22	Semi Culturable
	Sona Dighi	Vested	4.24	Culturable
	Chatraghati Dighi	Vested	28.15	Derelict
	Altadighi	Private	16.00	Culturable
Kusmandi	Mahipal Dighi	Vested	27.45	Culturable
		Total area (ha)	409.52	

Table 45: Block-wise List of Big Water Bodies of Uttar Dinajpur District

Sl.No.	Name of Block	Mouza	Water Area in Acre
1.	Chopra Block	Sona patila	5.81
2.		Teteswari	5.27
3.		Tepagachi (Bhotamari)	6.34
4.		Chpora (Chopradighi)	9.05
5.		Bhotamari	7.77
6.		Rajabhim (Nakanhar)	14.98
7.		Paschim Chutiakhor	3.71
8.		Ghirnigaon	3.64
9.		Borovilla	4.37
10.		Jiakharigachh	3.66
11.		Chitalghata	1.43
12.		Ghirinigaon Khas	1.98
13.		Ghirinigaon Khas	1.95
14.	Islampur Block	Sapnikla	7.5
15.		–do–	2.96
16.		–do–	2.97
17.		–do–	2.72
18.		Kulathan	2.23
19.		Sahatpur	2.24
20.		Ramgunj	1.97
21.		U/Khajagaon	1.72
22.		Agdimti	1.48
23.		Bolhanja	1.72
24.		Sahatpur	1.48
25.			
		Kuchila (Beel)	7.9
26.	Goalpokher-I	Ghhardhhappa	1.56
27.		Hanskunda	1.6
28.		Satbilla	2.4
29.		Baravilla	3.05
30.		Boggaon	1.82
31.		Bhindabari	2
32.		D/Basantapur	1.56
33.		Intia	15.05
34.		Panjipara	19.11
35.	Goalpokher-II	Hatwar	1
36.		Chathol	1.98

Contd...

Table 45–Contd...

Sl.No.	Name of Block	Mouza	Water Area in Acre
37.		Sagarpur	1.23
38.		Lohagunj	17.25
39.		Dhuabari	25.44
40.		Parul	5.42
41.		Parul	12.25
42.		Baggaon	11.23
43.		Barrren	12.61
44.		Patora	8.43
45.		Belan	13.07
46.		Belan	5.66
47.		Belan	5.25
48.	Hemtabad	Arjibahala	2.41
49.		Bamair	2.79
50.		Basudevpur	1.66
51.		Surangapur	3.16
52.		Sekhpur (Doba)	1.67
53.		Dehuchi	2.87
54.		Badhikobari(Doba)	1.9
55.		Chainagar	2.09
56.		Magra beel of Ghughurmal	8.43
57.		Naoda beel of Naoda	21.52
58.		Beel of Bhanail	12.78
59.	Kaliagunj	Taranga	1.85
60.		Mahinagar	2.97
61.		Mahinagar	2.43
62.		Mahinagar	1.96
63.		Mahinagar	2.54
64.		Sursha	2.26
65.		Bhuinhara	2.5
66.		Purba Durgapur	2.49
67.		Bhuinhara	5.32
68.		Dalimgaon	5.75
69.	Itahar	Indan	3
70.		Suran	3
71.		Sohar	1.8
72.		Malanchi	1.8

Table 45–Contd...

Sl.No.	Name of Block	Mouza	Water Area in Acre
73.		Gopibati	2.8
74.		Gopibati	1.6
75.		Gouripur	5.83
76.		Churaman	10.84
77.		Churaman	6.53
78.		Chandanpur	11.16
79.		Jagdal	5.58
80.		Keotal	19.6
81.		Rajgram	6.34
82.		Mahinagar	5.9
83.	Raigunj	Lohagara	5.56
84.		Rllgram	22.99
85.		Sarana	5.47
86.		Parial	10.69
87.		Dweepnagar	16.42
88.		Ital	22.17
89.		Matia	11.03
90.		Matia	61.39
91.		Sibpur	9.46
92.		Poaltar	8.93
93.		Raigunj	7.92
94.		Bishnupur	18.21
95.		Kotgaon	10
96.	Karandighi	Surta	13.52
97.		Surta	3.08
98.		Sushilapur	1.92
99.		Jughupur	2.7
100.		Jughupur	6.35
101.		Bhulki	2.48
102.		Gopalpur	3.26
103.		Gopalpur	3.87
104.		Baliamari	3.28
105.		Nengura	3.62
106.		Bajargaon	2.3
107.		Bajargaon	2.81
108.		Bajargaon	2.82

Contd...

Table 45–Contd...

Sl.No.	Name of Block	Mouza	Water Area in Acre
109.		Harsunda	4.74
110.		Deghalgaon	1.88
111.		Debattak Milik	2.59
112.		Debattak Milik	1.86
113.		Mahesh Bathna	1.82
114.		Karandighi	6.82
115.		Dhangaon	1.82
116.		Barhaneh of Bhagsala	9.92
117.		Chirandar of Bagindar	15.53
118.		Bazargaon of Bazargaon	240.3
119.		Bora	5.4
120.		Haria	5.66
121.		Hantara	21.19
122.		Mtiary	6.45
123.		Bagindar	19.5
124.		Ranigunj	20.35
125.		Dalkhola	9.85
126.		Kachnagachh	102.91
127.		Purba Mahespur	2.37
		Total Area (Acre):	1196.16

(*v*) Malda District

(*a*) Wetland of Malda

Being situated in the Gangetic flood plain there are many wetlands in southern West Bengal as compared to North Bengal. But the district of Malda is an exception. Out of 23 major wetlands in West Bengal 11 are situated in Malda district. River Ganga enters this district and while others flowing through this region are – Mahananda, Kalindi, Purnabhava and Pagla – which together create a lot of small and large water bodies (Figure 9). Floods are of frequent occurrence and during which more than half of the district is converted into a single large water body.

Although this district has a large number of wetlands, the local people are not much aware of the economic importance and ecological significance of this wealth. They may also serve as a source of drinking water in time of need if preserved property. Everywhere in Maldah district wetlands are being destroyed. Wetlands are essential for the ecological, sociological and economical balance of the district. Their disappearance may lead the disappearance life of animals, plants and humans.

The block-wise list of big water bodies of Malda district is depicted in Table 46.

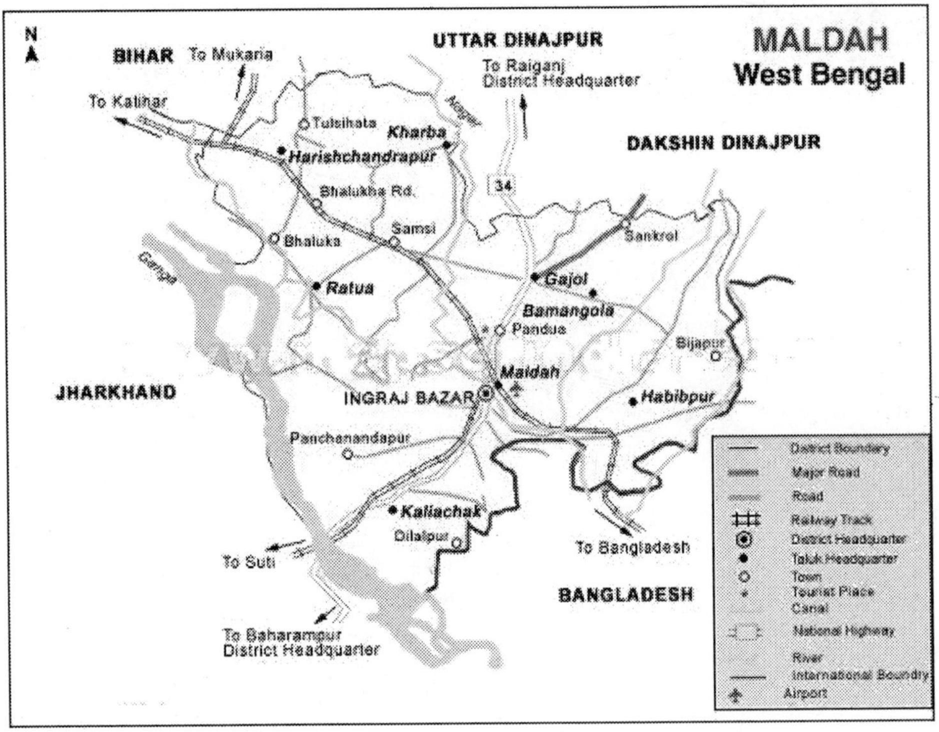

Source: www.mapsofindia.com

Figure 9: River Map of Malda District

Table 46: Block-wise List of Big Water Bodies of Malda District

Sl.No.	*Name of the Block*	*Name of the Water Body*	*Area (ha)*
1.	Kaliachak-1	Taraghati Beel	8.41
2.		Marabhagirathe beel	6.86
3.		Maramasi beel	12.4
4.		Pagla beel	8.1
5.		Jalua basa beel	8.48
6.		Chatta azgbi dhap beel	11.11
7.		Trimohona beel	18.1
8.	Kaliachak–II	Babla dhap beel	21.2
9.		Gurnamani beel	47.2
10.		Garutala barabadha beel	5.52
11.		Azgubi dhap beel	22.94
12.		Sripur dalua grame beel	8

Contd...

Table 46–Contd...

Sl.No.	Name of the Block	Name of the Water Body	Area (ha)
13.		Barkul dhap beel	18.84
14.		Nimazale beel	10.3
15.		Sattary beel	49.02
16.	Kaliachak–III	Sukpara beel	12
17.		Golap gunj dhap beel	9.46
18.		Sobdhal pur beel	30
19.		Deona pur beel	65.18
20.		Surja pur beel	58
21.		Koradhap beel	9.36
22.		Charudhap beel	22.69
23.		Sovanayar beel	13.04
24.	English Bazar	Doha Kandar beel	5.07
25.		Veoa beel	5.85
26.		Rumari beel	9.3
27.		Nandar beel	26.85
28.		Bahagola beel	6.98
29.		Beel	9.24
30.		Bital beel	6.78
31.		Kalimata beel	9.64
32.		Amriti beel	10.16
33.		Gabgachhi beel-1	32.52
34.		Gabgachhi beel-2	31.28
35.		Beel	30.2
36.		Mora Bhagirathi beel	36.56
37.	Old Malda	Chata daba beel	6.88
38.		Kalna beel	9.61
39.		Sona kanehan beel	31.18
40.		Bekey beel	5.45
41.		Chatra beel	2.03
42.		Naluary beel	59.32
43.		Koari beel	59.32
44.		Jalkar Bithan beel	224.4
45.		–do–	5.62
46.		–do–	6.33
47.		–do–	2.8

Contd...

Table 46–Contd...

Sl.No.	Name of the Block	Name of the Water Body	Area (ha)
48.		Jalkarbithan Beel	14.13
49.		–do–	30.33
50.		–do–	10.11
51.		–do–	17.15
52.		Matikata beel-I	7.47
53.		Matikata beel-II	8.71
54.		Mamud beel	7.78
55.		–do–	5.15
56.		–do–	3.04
57.		Santipur beel	18.22
58.		Sona beel	3.76
59.		Mora nadi beel	2.88
60.		Chamar kale beel	12.97
61.		Chatra beel	4.74
62.		Ranju beel	5.49
63.		Putul beel	22.68
64.		Ghahana beel	7.54
65.	Gozal	Lohachora beel	20
66.		Purba beel Ahora	36
67.		–do–	2.08
68.		Kaccha Kati beel	2.5
69.		Mahidurga beel	31.12
70.		Sumari beel	12
71.		Chardhar beel	17.04
72.		Hatia beel	10.69
73.		Projapoti beel	6.06
74.		Singhar beel	4.17
75.		Basu beel(1)	11.55
76.		Beel parla	12
77.		Beel Kanehan	2.13
78.		Dighali beel	1.42
79.		Ververia beel	6.89
80.		Berabhon beel	8.38
81.		Kanehan beel	13.38
82.		Kole beel	2.48
83.		Dahedar beel	7.08
84.		Dharmadanga beel	7.31

Contd...

Table 46–Contd...

Sl.No.	Name of the Block	Name of the Water Body	Area (ha)
85.		Gokhria beel	3.94
86.		Kandiar beel	6.84
87.		Saluka beel	3.71
88.		Altar beel	7.17
89.		Mara beel	3.4
90.		Parlla beel	3.63
91.		Mora nadi beel	5
92.		–do–	8.38
93.		Khaghsura beel	3.98
94.	Bamongala	Dokram beel	6
95.		Tehattai beel	8
96.		Charkale beel	6.22
97.		Palta beel	8
98.		Kodal kati beel	7.6
99.		Chatra beel	12
100.		Mora nadi beel	12.1
101.		Sankar beel	23.95
102.		Nandina beel	6.59
103.		–do–	5.91
104.	Habibpur	Siruli beel	5.6
105.		Dhal mara beel	8.82
106.		Baman beel	29.31
107.		Nandaghar beel	7.78
108.		Vandaly beel	6
109.		Mandala beel-1	23.05
110.		Mandala beel-II	6.02
111.		Pither beel	11.85
112.		Bhavani beel	7.75
113.		Chakli beel	10.8
114.		Hirmari beel	9.8
115.		Kuehiamera beel	13.38
116.		–do–	9.58
117.		Mahadeb pur	6.76
118.		–do–	16.64
119.		Alanda beel	7.95
120.		Thanar beel	10.12

Contd...

Table 46–Contd...

Sl.No.	Name of the Block	Name of the Water Body	Area (ha)
121.		Dhuku beel	11.04
122.		Subarni beel	6.6
123.		Digali beel	23.57
124.		Dhari duba beel	29.86
125.		Dharsuna beel	6.53
126.		–do–	8.05
127.		Harischandrapur	23.42
128.		Sriram beel	9.59
129.		Ramdoal beel	8.59
130.		Dhamas beel	21.54
131.		Maranadi beel	11.6
132.		–do–	21.57
133.		Mohina beel (Srikrisnapur)	5.44
134.		Buridara beel	6.13
135.		Rikon beel	10.64
136.		Karał beel	6.93
137.		Adhar kata beel	4.12
138.		Bokla beel	11.46
139.		Mohona beel (Palharrabsha)	10.87
140.	Ratua -I	Talbona beel	22.85
141.		Haripur cob beel	19.64
142.		Rangamatia beel	35.33
143.		Baramasia beel I	14.36
144.		Baramasia beel II	16.18
145.		Bizri vitta beel	12.76
146.		Kuji beel	24.75
147.		Purnivisa beel	18.29
148.		Bahati beel	22.98
149.		Gulijana beel	42.33
150.	Ratua -II	Bora villa beel	15.19
151.		Nayacate beel	11.23
152.		Pabiar beel	15.92
153.		Kaul Kand beel	7.22
154.		Khulaha-ccb	46.2
155.		Kala Pahar beel	10.7
156.		Golbaka beel	30.12

Contd...

Table 46–Contd...

Sl.No.	Name of the Block	Name of the Water Body	Area (ha)
157.		Lamba baku beel	10.68
158.		Bhoga jalkar beel	42.3
159.		Mechua beel	22
160.		–do–	13.42
161.		Gahular beel	12.31
162.		Kamdal beel	24.9
163.		Bhyaryari beel	10.45
164.		Cha Pair beel	14.37
165.	Manichak	Atisaiya beel	11.5
166.		Barabadha beel	14.38
167.		Puline tala beel	44.08
168.		Bahir cala beel	4.74
169.		Kalithana Baracate tatipita beel	9.64
170.		Amtala beel	22.32
171.		Narayan pur beel	32.33
172.		Laskar pur Madia beel	20.52
173.		Laskarpur beel	27.28
174.		Hilsa mani beel	23.02
175.		Bochahi beel	9.84
176.		Baramasia beel	6.8
177.		Galaghat beel	23.32
178.	Chanchal -1	Chalmura beel	17.38
179.		Siali beel	6.82
180.		Domai beel	5.12
181.		Chandra beel	80.12
182.		Guchimari beel	15
183.		–do–	8.88
184.	Chanchal-II	Barasingha beel	21.46
185.		Patsal beel	8.24
186.		Chapra beel	9.44
187.		Chandrasahi beel	70.7
188.		Sahajob beel	8.21
189.		Surial beel	8.6
190.		Ghoradoha beel	49.8
191.		–do–	8.04
192.		Sarabani beel	26.36
193.		Singin beel	8

Contd...

Table 46–Contd...

Sl.No.	Name of the Block	Name of the Water Body	Area (ha)
194.		Singin beel	9.76
195.		Sinyra beel	12.97
196.		Sonakhari beel	22.68
197.	Harishchandra Pur –I	Moranadi beel	15.31
198.		Hazartakiar beel	19.8
199.		Chang beel	18.05
200.		Ashideb beel	10.84
201.		Kopaichandi beel	13.53
202.	Harishchandra Pur –II	Dumuria beel	22.37
203.		Sialy beel	21.4
204.		Pajuar beel	10
205.		Kochuar beel	22.97
206.		Khasimari beel	19.2
207.		Vedul beel	13
208.		Dhekul beel	24
209.		Mutna beel	9.11
210.		Khankhore beel	18.26
211.		Gitarmoni beel	9.61
212.		Sanak beel	12.04
213.		Kaia beel	10.75
214.		Matal beel	8.41
215.		Patibona beel	17.45
216.		Manna beel	27.31
		Total Area (ha)	3564.43

(*b*) Makhana Cultivation

Makhana is the popped expanded kernel of the gorgon nut (*Euryale ferox*) and is used as a delicious food (Plate 12). Presently, it is sorted, graded and used as an ingredient of various ready-to-eat products. Makhana requires further processing to get some value-added products of minimum volume. To develop any kind of product and mechanized system for their production, physical properties are required. Furthermore, since makhana is a seasonal and regional crop its storage conditions are also needed for keeping it appropriately at processing centres. Physical properties of makhana at moisture contents ranging from 5 to 20 per cent (dry basis) and equilibrium moisture content at relative humidities ranging from 11·2 to 88 per cent at temperature 30°C were determined using standard techniques. Physical properties were found to be: test weight (weight of 1000 makhana) 286 to 384 g; bulk density 56.5–84.6 kg/m^3; particle density 105–240.6 kg/m^3; porosity 29.4–48.9 per cent; angle

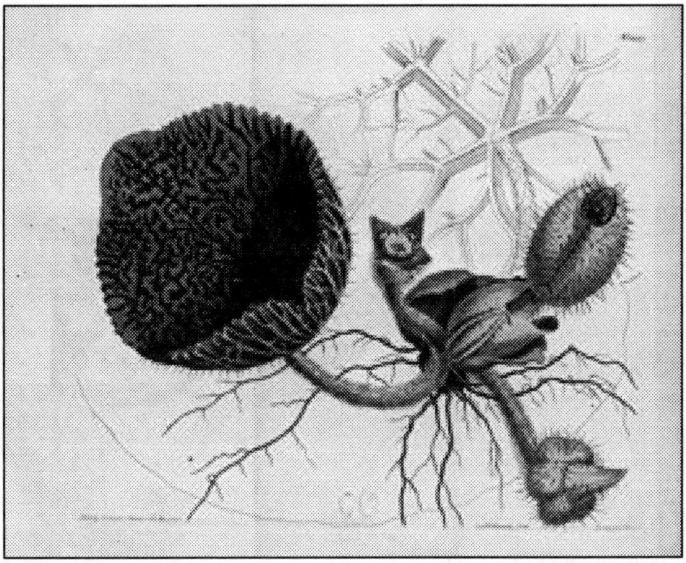

***Source*: Internet**

Plate 12: Illustration of *Euryale ferox* from *Curtis's Botanical Magazine* (1812)

of repose 33–35.6°; and static coefficient of friction 0.596–0.82 and 0.493–0.684 on galvanized iron and stainless steel, respectively. The equilibrium moisture content of makhana was found to be between 11.5 and 58.9 per cent (dry basis) within the ranges of variables studied. (Ref.: Silsoe Research Institute)

Euryale ferox (also known as fox nut, foxnut, makhana, or gorgon plant) is the only species in the genus *Euryale*. It is a flowering plant classified in the water lily family, Nymphaeaceae, although it is occasionally regarded as a distinct family Euryalaceae. *Euryale* is an annual plant native to eastern Asia, and is found from India to Korea and Japan, as well as parts of eastern Russia. It grows in water, producing bright purple flowers. The leaves are large and round, often more than a meter (3 feet) across, with a leaf stalk attached in the center of the lower surface. The underside of the leaf is purplish, while the upper surface is green. The leaves have a quilted texture, although the stems, flowers, and leaves which float on the surface are covered in sharp prickles. Other leaves are submerged.

Kingdom:	Plantae
Division:	Magnoliophyta
Class:	Magnoliopsida
Order:	Nymphaeales
Family:	Nymphaeaceae
Genus:	*Euryale* Salisb., 1805
Species:	*E. ferox*
Binomial name	*Euryale ferox* (Salisb., 1805)

The plant produces starchy white seeds, and the seeds are edible. The plant is cultivated for its seeds (Plate 13) in lowland ponds in India, China, and Japan. The Chinese have cultivated the plant for over 3000 years. More than 96,000 hectares of Bihar, India, were set aside for cultivation of *Euryale* in 1990-1991. The plant does best in locations with hot, dry summers and cold winters. Seeds are collected in the late summer and early autumn, and may be eaten raw or cooked. Its edible seeds (Plate 14) are used in traditional Chinese medicine, where they are often cooked in soups along with other ingredients, and believed to strengthen male potency and retard aging. In India, particularly in the northern and western parts of the country, *Euryale ferox* seeds are often roasted or fried, which causes them to pop like popcorn. These are then eaten, often with a sprinkling of oil and spices.

Source: Internet

Plate 13: A Pond of Cultivated *Euryale*

Source: Internet

Plate 14: Gordon *Euryale* Seed

The name *Euryale* comes from the mythical Greek Gorgon by the same name. The Soviet Union issued a postage stamp featuring this species.

Latin

Semen Euryales.

Origin

This herb is the endocarp of ripe seed of *Euryale ferox* Salisb., an annual water plant of the family Euryalaceae. Native to East Asia, the plant is grown in ponds and lakes in lowland in China, Japan and India.

The plant grows to about 1 m by 3 m. The flowers are hermaphrodite (have both male and female organs). The plant cannot grow in the shade. It can grow in water.

In China, Gordon euryale is produced mainly in Hunan, Jiangsu, Anhui, Shandong and other provinces. During August and September every year, it is collected, dried in the sunlight after getting rid of shell, epicarp and the brown internal shell, and used unprepared or stir-baked and crushed. This is also called Foxnut.

Properties

Sweet and astringent in flavor, mild in nature, it is related to the spleen and kidney channels.

Functions

Astringes intestines to arrest diarrhoea, tonifies the kidney to preserve kidney essence and cures leukorrhagia. Gordon euryale seed is a sweet and sour astringent herb that acts as a tonic for the kidney and the spleen.

Applications

Tonigying the kidney to arrest seminal emission:

For kidney-deficiency manifested as spontaneous emission and spermatorrhea, it is used with herbs for reinforcing the kidney to preserve essence.

Removing dampness to arrest abnormal leukorrhea:

For leukorrhagia due to insufficiency of the spleen with over-abundance of dampness, it is used with herbs for invigorating the spleen to arrest leucorrhagia, such as radix codonopsis pilosulae, rhizoma atractylodis, rhizoma atractylodis macrocephalae, rhizoma dioscoreae, etc.

In treating long-lasting diarrhoea due to deficiency of the spleen, the herb is often used in combination with radix astragali seu hedysani, radix codonopsis pilosulae, rhizoma atractylodis macrocephalae, rhizoma dioscoreae, poria, etc.

Modern Researches

Gordon Euryale seed is analgesic (insensitizes pain) and aphrodisiac (arouse sexual desire). It is taken internally in the treatment of chronic diarrhoea, vaginal discharge, kidney weakness associated with frequent urination, impotence, premature and involuntary ejaculation and nocturnal emissions.

Gordon Euryale leaf is used in cases of difficult parturition (process of giving birth).

All parts of the plant are considered to be astringent, deobstruent (removes obstructions) and tonic.

Fox nut or gorgon nut (*Euryale ferox* – Family Nymphaeaceae), popularly known as Makhana, has been widely used in traditional oriental medicine to cure a variety of diseases including kidney problems, chronic diarrhoea, excessive leucorrhea and hypofunction of the spleen. Based on the recent studies revealing antioxidant activities of Euryale ferox and its glucosides composition, sought to be determined if Euryale ferox seeds (Makhana) could reduce myocardial ischemic reperfusion injury. The hearts of the Makhana treated rats were resistant to ischemic reperfusion injury as evidenced by their improved post-ischemic ventricular function and reduced myocardial infarct size. The Makhana-treated hearts had increased amounts of thioredoxin-1 (Trx–1) and thioredoxin-related protein-32 (TRP32) compared to the control hearts. Different studies demonstrate cardioprotective properties of Makhana and suggest that such cardio-protective properties may be linked with the ability of makhana to induce TRP32 and Trx-1 proteins.

Gorgon nuts or Dalmakhna (*Euryale ferox*) is popularly known as Makhana is cultivated in the low lands where minimum 1.5 ft to 3 ft water depth is required.

Plant

Makhana looks like as lotus plant. Broads, round shaped leave, thorny stalk (Plate 15). Leaves covered upper surface of the water body. After maturation flowers blossom. Each flower consists of 20-30 blue/violet/red petals and it is complex in nature. Fruit is formed from the flower. Each fruit contain 20-50 black colored seeds. The fruit gradually bent and faced towards pond bed. In due course fruit split and the seeds begin to float in to the water. After few days the seeds absorb water and sink to the pond bed. Mean while plants begin to rot.

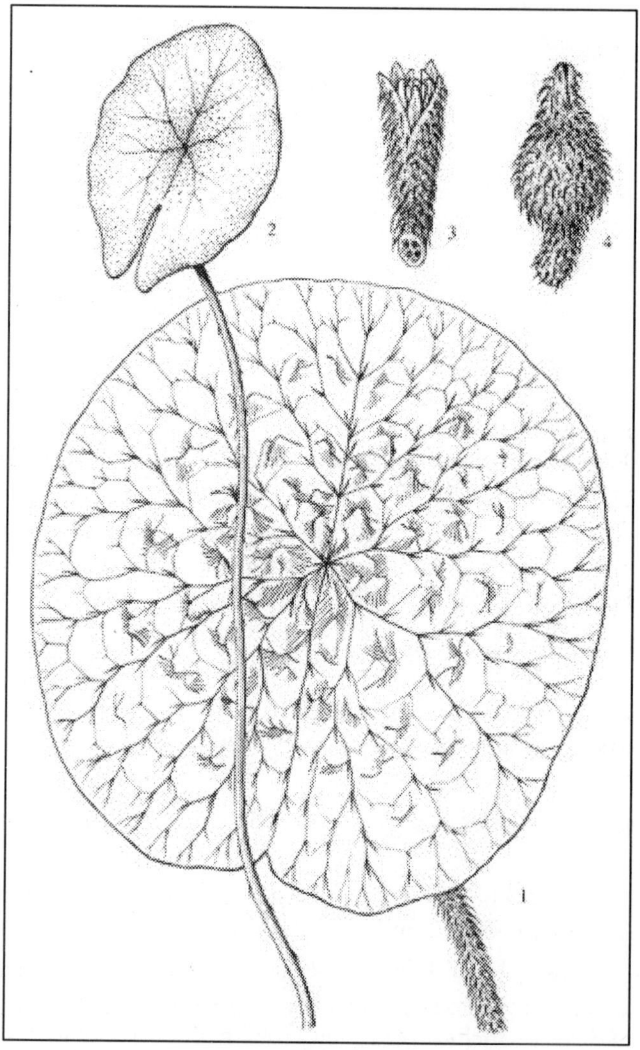

Plate 15: *Euryale ferox*
1: Leaf; 2: Young leaf; 3: Flower; 4: Fruit.

Seed

It is round in shape and black in color and made up of fleshy white pulp. It looks like as sunflower seed. When all Makhana Plants are rotten, seeds are collected by experienced labour for this purpose by a conical shaped bamboo made basket.

Food Value

It is nutritious and highly protinous food mainly used in baby food as it is easily digestible.

Economic Value

The seeds are fried to make "Khai" which is good in taste and has a great demand in the national and international market. It also has some medicinal value.

Makhana Production in Malda district of West Bengal

Production

20-25 quintals per ha.

Cost

Raw seeds are sold @ Rs. 50.00/k.g. whereas "Khai" is sold @ Rs. 120-150/k.g.

Area of Cultivation in Malda District of West Bengal

Under Harischandra Pur-I about 1200 ha and Harischandra Pur-II about 1500 ha.

Fishery Point of View

10 per cent -20 per cent Makhana cultivated water body are covered as fishery mainly air breathing fishes are stocked. Makhana cultivators are advised to spare some area in the middle of the water body and release IMC (Except grass Carp and Cyprenious Carp) in the month of June-July. As a result more than 700 fish farmers are engaged in fishery under Harischandra Pur-I and Harischandra Pur II Block, Malda.

Yield

@ 500 k.g./ha fishes are yielded from Makhana Water Body.

Weather

Makhana grows well in warm weather. Mild rain enhances its growth whereas heavy rain reduces. Rain with hail is dangerous for the crop.

Soil

It can be cultivated in any water body. Generally less deep wetlands having a layer of 15-20 cm of clay and silt is ideal for Makhana cultivation.

Process of Cultivation

- ✰ Preparation of the seed bed using organic manure after 2-3 times of cultivation.
- ✰ Seeds are spread at the ratio of 5kg/bigha.
- ✰ Preparation of the main land with 2-3 times of cultivation. Keep water upto 2 ft depth over the land
- ✰ Seedlings are collected from the seed bed (1 ft long with 2-3 leaves) and planted in the main land.

Use of Manure

Generally no manure is needed as the rotten weeds naturally make the organic manure. Otherwise, according to the fertility of the wetland 1.5-2 ton/bigha organic manure can be used.

Collection and Processing

The seeds are collected from the wetlands and dried in sun. Then it is fried and hammered to make "Khai".

Preservation of Seed

After total drying in sun, the seeds are packed in bags and stored in clean place. Seeds can be kept immersed in a water body packed with nylon net for next year cultivation.

Diseases and Control

(*i*) *Pamri Poka*: It eats up different parts of the roots and damages it. To control it is advised not to cultivate in the same land repeatedly.

(*ii*) *Noli Poka*: Mature one lays egg on the leaves. The larva eats up the green parts of leaves and flowers. Rubbing of jute ropes dipped in kerosene over the infected leaves can help to control the pest.

(*iii*) *Jab Poka*: It sucks the juice of leaves and soft parts of the plant. As a result plant rotten and died. Spraying of water over the infected leaves can help to control it.

(*iv*) *Rotten Disease*: The roots and stems get rotten slowly and the plant dies. To control this disease the diseased plant should be removed immediately. It is advised to keep the water clean.

In Brief

- ☆ *Time of Cultivation*: Between 15th November to 15th December
- ☆ *Time of Seed Collection*: From 15th August–October last
- ☆ *Time of Flowering*: Middle April-November
- ☆ *Time of Fruit Arrival*: Middle of May

Economics of Makhana cultivation (Per Bigha)

1.	Preparation of land for cultivation (Clearing of weed, to cultivate etc.) 5 person x 40/-	.... Rs. 200.00
2.	Cost of Seed (5 kg x 30/-)	.... Rs. 150.00
3.	Pesticide (Aquatic insects, fishes)	.... Rs. 50.00
4.	Weeding hook, (5 person x 40/-)	.... Rs. 200.00
5.	Harvesting and etc.	.... Rs. 1400.00
	Total	.... Rs. 2000.00

Total income: Rs. 300.00/Bigha x Rs. 30.00 = Rs. 9,000.00

Net Profit: (Rs. 9,000.00–Rs. 2,000.00)/bigha = Rs. 7,000.00

Chapter 3

Jhora Fishery in Hill Area of Darjeeling District

The *"Jhora"* is known as "Spring water" in Nepali language. The fish culture pond that is fed by the perennial spring water or Jhora is commonly known as "Jhora Fishery". Generally an area having 1500 sq. ft. is called one unit Jhora pond with probable dimension of 50 ft. length and 30 ft. breadth. The slope of the bottom should be towards outlet at about 0.2 per cent. Depth of the *Jhora* pond varies between 0.8 and 1.0 m. There should not be any trees around embankment as the roots can cause water seepage. The ideal *Jhora* ponds should have a gentle continuous flow of water.

Jhora exists in two forms *viz.* (i) perennial *Jhora* and (ii) seasonal *Jhora.*

(*i*) *Perennial Jhora:* Water is available throughout the year in perennial Jhora or Spring.

(*ii*) *Seasonal Jhora*: Water is available only during monsoon season and some part of the year in seasonal *Jhora.*

Both the above *Jhoras* are available in Darjeeling district, West Bengal where torrential water flows through deep gorges with steep wooded sides. Their beds are covered with boulders, rock and stones over which water flow in series of water falls, cascades and rapid with pools of various size here and there. Aquatic fauna consists of earth worms, round worms, insect larvae, larvae of simuliidac mostly chironomids, smaller crustaceans, may fly larvae, mosquito larvae, corixid bugs, nymphs, pleoptera. Diatoms predominate as attached organisms. The desmids, closterium and cosmariam also exist. During summer it is snow and spring fed, while in monsoon the streams are heavily fed with rain water. A maximum depth exists before monsoon in small streams at 3 ft. and in large streams at 10 ft. Generally four types of Jhora ponds are found.

(*i*) General/Normal pond with muddy embankment all around,

(*ii*) Mud bottom pond with only muddy bottom and embankments in all the four sides are made up of concrete,

(*iii*) Cemented bottom pond where the pond bottom is covered with cement mortar mixture,

(*iv*) Total Cemented Pond where the whole pond both bottom and walls (embankments) are constructed with rods, stones/gravel//bricks etc.

Among the four types, any person having a piece of free space adjoining his dwelling house and having facility of perennial Jhora in the vicinity may easily decide for conversion of the land into Jhora Pond and start Jhora Fishery. The conceived idea and blending of technology base, compiled with the peoples participation is linked to all economic sector(s) in the field of Agriculture and Allied sectors in general and Fisheries/Aquaculture in specific, is no way exception to it. Culture of such ideas has called for harnessing the upland fisheries resource base along the Darjeeling Hills.

3.1. What is *Jhora* Fishery?

The fish culture pond that is fed by the perennial spring water or *Jhora* is commonly known as "*Jhora* Fishery". Fish culture practices in the hill areas have been reported to be since time immemorial though not on a commercial scale. The people used to collect fingerlings from the rivers, streams and keep them in their backyard or the kitchen pond. These types of fish culture were done more as hobby and decorative purpose rather than financial gains. Demand of fish as food to meet the protein requirement of the hill people were thought by the Department of Fisheries, Government of West Bengal and steps towards popularization of *Jhora* fishery was taken up through scientific management practices. Unit area of the *Jhora* Fishery is termed as one unit, half unit, etc. Generally an area having 1500 sq ft is called one unit *Jhora* with probable dimension of 50 ft. length and 30 ft. breadth. It signifies culture of fish in small ponds fed by the perennial or seasonal mountain spring.

3.2. Concept of *Jhora* Fishery

The main concept in this lies in storing *Jhora* (stream) water (Plate 16) in a small ditch, of irregular size by constructing barricade with the help of earth, sand and stone. The dissolve oxygen (D.O.) content in this water is high. Cold tolerant Fish species from the plain land, after due acclimatization, are generally stocked and cultured. Generally water retains throughout the year or part of the year in the *Jhora*. Accruing the benefit of continues flow of the feeder *Jhora*, a constant water level is maintained in the *Jhora* Pond and this facilitates culture throughout the year or part of the year with high stocked biomass.

3.3. Prospects of *Jhora* Fishery

Jhora ponds could be a great resource for fishery in the hills of Darjeeling district, if properly managed. In most of the *Jhora* ponds fishes are cultured, without proper scientific management protocols. This has resulted in poor production and meager

Plate 16: View of Jhora (Stream)

gain in revenue. Further, a number of constraints are coming in the way which may be attributed to non-maintenance of ideal depth and seepage problems, poor soil and water quality and non-availability of proper feed, which ultimately leads to either poor nutrient level or eutrophication.

3.4. History of *Jhora* Fish Culture

The concept of *Jhora* pond fish culture dates back to long ago. Till 1980, there was no incidence of organized fish culture approach in the Darjeeling Hill areas excepting capture and angling from the Hilly rivers. In the year 1981, the Department of Fisheries, Government of West Bengal, attempted fish culture in small excavated ponds fed by *Jhora* water in Kalimpong Sub-Division of Darjeeling district by setting up nine small units (13.0 m^2 each) as demonstration centres at different altitudes to study the growth of fishes. Economically viable fish species that thrive well in the agro-climatic conditions of Kalimpong area of Darjeeling district were selected. After managing culture operation for nine months, table fishes having a total weight of 100-120 Kg (7.5-9.0 tons/ha) were harvested from each pond which is considered quite high in hill conditions. These encouraging results prompted the State Fisheries Department to promote this kind of aquaculture system on a large scale. The Government provides 50 per cent subsidy assistance for the beneficiaries. So far more than 4193 (Table 47) Jhora fishery units have been constructed assisting under Fish Farmers Development Agency (FFDA) programmes.

Table 47: Block-wise Jhora Fishery Already Commissioned

Sl.No.	Name of the Blocks	Total Number of Jhora Fishery Units
1.	Darjeeling-Phulbazar	790
2.	Rangli-Rangliot	587
3.	Jor.-Sukhiapokhari	369
4.	Kalimpong-I	828
5.	Kalimpong-II	451
6.	Gorubathan	369
7.	Kurseong	447
8.	Mirik	352
	Total No.	4193

The hill fisheries in West Bengal have received very little attention though scope exists for small-scale development. The Darjeeling District in its four sub-divisions, *viz.*, Kalimpong, Kurseong, Darjeeling and Siliguri possess fish cultivable area of 2000 ha and riverine resources of over 4000 ha. Excepting a few individuals who opted as hobby, fish culture was not practiced as a commercial venture by any one in the hills till 1983. Little thought was given that the ponds could easily be constructed on broken waste land, through utilization of waste land, marshy land, derelict water meadows, gullies and ravines for fish culture. Depression on the ground, a succession of shallow little valleys or ravines, muddy water holes are the source of excellent ponds and by skilful use of the levels in undulating lands, though ponds can be constructed without much expense of labour. A number of small, shallow ponds are much preferred than having one or two large ones. From the economic point of view, it is better to construct ponds which require minimum excavation and artificial banking.

3.5. Basic Criteria for Site Selection of *Jhora* Fishery

Like any aquaculture system it is crucial to choose a suitable site for *Jhora* fishery. The main considerations in proper site selection are topography, water supply and soil quality. The ideal location for a *Jhora* fishpond is a flat land or nearly so. In hill areas whenever ponds are built on sloping ground they are mostly irregular in shape. The best site for *Jhora* ponds must possess small stable drainage system where the possibilities of flooding and landslide are remote. There are hundreds of perennial hill springs in the Darjeeling hill. It is equally necessary to have a sufficient good quality of water supply for the *Jhora* ponds. So, the appropriate site of a *Jhora* pond must be located near a perennial *Jhora* or stream from where water is channeled into it. Alternative arrangement of water supply should be made for dry weather. However, ponds should be on lower elevated and so that the surface runoff over the surrounding catchment area can be accumulated in the ponds. Soil having the clayey-loamy texture is suitable for better retention or water in a *Jhora* pond since these have the maximum water retentively and can be easily compacted and made leak proof. Another consideration is that it is desirable to construct the pond on a land around a household or within the vegetable garden (commonly known kitchen ponds) for convenience of management.

3.6. Management of Jhora Ponds

(*i*) Pond Construction

A plot having least chances of seepage and horizontal percolation is suitable for constructing the pond. If the area is having seepage, it can be prevented through promoting growth of agri-mat lining to prevent seepage. The central part of the pond is generally made somewhat deeper but less than 1 m depth and having a gradual raising bottom culminating in shallow marginal areas. Plantation of small herbs around the pond helps to check soil erosion. Presently, mason tanks are preferred by *Jhora* beneficiaries so that, the cost of maintenance on earthen ponds is reduced.

(*ii*) Shape, Size and Depth of Jhora Pond

Fish pond constructed in the land around the house or within the vegetable garden of the farmer near a perennial *Jhora* or stream is conducive for management. The shape of the pond may be round, oval, square or rectangular with an approximate area of about 15–20 sq. meters. However, many larger ponds are also being constructed. Depth of the *Jhora* pond varying between 0.8 and 1.0 m. Of course no specific size can be quoted for hill ponds due to the sloping and rugged topography. The minimum size should be 5 sq. metres. Small ponds are easier and quicker to harvest, drained and refill to treat diseased. If possible, the pond should be rectangular (length: breadth 3:1). The breadth and length should not be more than 30 to 40m and 40m respectively. The corners should be rounded so that the fishes may not escape, and congregate in the corners. Normally depth of the pond varies according to the species to be cultured. The larger fish needs greater depth and an average depth of 2.5 to 5 ft is recommended. At the bank it should never be less than 19 inch in order to avoid invading emergent vegetation. The slope of the bottom should be towards outlet at about 0.2 per cent,

(*iii*) Embankment of Jhora Pond

The embankment should be strong enough to be stable in all weather and not liable to collapse during heavy rain. The two main qualities are solidity and water retention capacity. The slope of embankment may be 2:1 or 3:1. The width may vary from 1.0-1.5m. The embankment must be 12 inch above surface of water for small pond and 20 inch for large pond. Very shallow and too deep ponds are not conducive for fish culture in the hills. Too deep ponds are not preferable because of low water temperature whereas, in very shallow ponds the fishes are likely to be vulnerable to birds of prey. There should not be any trees around embankment as the roots can cause water seepage.

(*iv*) Water Inlet and Bypass Drainage

Management of water supply is important for raising fishes in *Jhoras.* Both overflow and complete drying should be avoided. Water channelized through passage must not enter the pond directly where it should fall from a higher level into the pond to prevent fishes from escaping into the stream. Further, flow of water from a higher level also facilitates aeration. Proper arrangement to check flooding during rains is also essential. The inlet and outlet must have a wire net or bamboo grating to check fishes from escaping to the outside. The ideal *Jhora* ponds should have a gentle continuous flow of water. The pond bottom must be covered with a network of drain or evacuation ditches of "fish bone design". The erosion of pond embankment can be controlled by laying stones or grass along the base of water flowing channel at the inlet points.

The three kinds of springs are generally found namely, *Rheocrene, Limnocrene* and *Helocrene. Rheocrene* are running springs in which on emergence from rocks, the water flows down the valley or towards it. *Limnocrene* springs are located in depressions which are first filled by the emerging water. The spill-over water follows its own course. In *Helocrene* springs the emerging water seeps through a layer of earth converting it into a marsh. The *Jhora* are basically of the *Rheocrene* and *Limnocrene* types. Water from *Rheocrene* and *Limnocrene* are clear and the temperature more or less stable and therefore, considered for Jhora fishery. Water quality studies in the *Jhora* ponds have shown high dissolved oxygen content and a very low nutrient level.

3.7. Types of *Jhora* Pond

Four types of *Jhora* ponds (Plates 17–22) are in vogue in the hill areas of Darjeeling:

(*i*) General/Normal Pond

The land areas of nearby to the house of the farmers are considered suitable for construction of the *Jhora* and are excavated for preparation of such pond. The overall depth of the pond is maintained about 2–4ft with muddy embankment all around. Since the nature of soil in these area are gravelly/sandy with poor compaction, embankments are prepared very carefully by mixing clay soil with the native soil to make it strong enough so that during monsoon the same do not get washed away. Before final compaction of the embankment, rubber tube of 1"diametre is connected with the nearby perennial *Jhora* in such a manner that the *Jhora* water can get entry into the pond directly with the gravity flow from the top in a corner. Simultaneously, similar type of pipe is also fixed at a fixed depth so that the excess water can bet out. Proper care is taken to ensure that flow of inlet and outlet is in harmony and the desired depth is always maintained.

Plates 17–22: Different Types of Jhora Ponds in Darjeeling District

Plate 17

Plate 18

Plate 19

Plate 20

Plate 21

Plate 22

(*ii*) Mud Bottom Pond

In this type of pond only bottom is made of soil/mud (Plates 23–24). The embankments in all the four sides are made up of concrete with cement morter. Due to its sandy nature the bottom soil has less water retention capacity. To minimize percolation loss and to maximize retention of water, some quantity of raw cow dung/ pig dung or any other decomposed organic manure is spread over the bottom soil.

(*iii*) Cemented Bottom Pond

In this type where the soil is highly gravelly/sandy and highly percolating, the bottom is made semi-concrete. In the similar manner as described in above, the pond is excavated and the pond bottom is covered with cement mortar mixture (rubble in cement–1 : 4, the safe permissible load of which is found to be 85 tones/m^2) of 4-6 inches deep, over which again excavated soil of 4-6 inches deep is laid to give the pond a natural simulation to assist in biological metabolite decomposition. But in all cases the depth is maintained as in previous types.

(*iv*) Cemented Jhora Pond

In this type the whole pond both bottom and walls (embankments) are constructed with rods, stones/gravel/bricks etc (Plates 25–30). Hence, this one is highly cost-intensive, without sufficient subsidized assistance this may not be affordable to the poor beneficiary. However, such type of pond is being advocated under FFDA-DGHC subsidy scheme. Among the four types, any body having a piece of free space adjoining his dwelling house and having facility of perennial *Jhora* in the vicinity may easily decide for conversion of the land into *Jhora* Pond and start *Jhora* Fishery.

3.8. Pre Stocking Management

Soil and water management is an essential step for optimizing fish production. Natural productivity of a pond can be greatly enhanced by the use of manures and fertilizers, which provide essential nutrients for aquatic biota serving either directly of indirectly, through involved ecosystem, as food of fishes.

(*i*) Liming

Aiming at fast mineralization of organic matter and for prophylactic reasons, use of quick lime @ 700 kg/ha (9 kg/unit) is recommended. Initial dose @ 100 kg/ha (1.394 kg/unit) may be applied one week earlier to stocking the pond and the rest in equal monthly installments. Quick lime need to be dissolved in water, allowed cooling down and broadcasting uniformly over the pond surface during morning hours. Besides, use of lime @ 100-200 kg/ha (1.394–2.788kg/unit) in one dose may be necessary when fishes show distressed conditions.

(*ii*) Manuring

Organic manures like cow dung, pig dung and poultry droplets and tea garden waste are used by the farmers for fertilising their ponds. In order to reduce the culture cost, they also use the mixture of cow dung: pig dung: poultry manure: tea garden waste in the ratio of 1: 1:1:1 and apply those in installments. A combination of organic

Plates 23–24: Jhora Pond with Mud Bottom and Cemented Walls

Plate 23

Plate 24

Plate 25: Pond with Cemented Bottom

Plate 26: View of Fish Seed Farm at Reyang

Plates 27–30: Aerial View of Cemented Jhora Fish Ponds

Plate 27

Plate 28

Plate 29

Plate 30

manure and inorganic fertilizer is considered more effective than either this alone. The fertilization schedule is prepared on the basis of fertility status of soil, which is ascertained by chemical analysis.

Initial manuring with organic manure @ 20 per cent of the total requirement is done 15 days prior to stocking. The remaining 80 per cent of the organic manure and the total quantity of inorganic fertilizer may be applied in split up 11 equal monthly doses during the rearing period. Manuring/fertilization need to be suspended in adverse ecological conditions. Selection of nitrogenous fertilizers are dependent on the status of soil pH *i.e.* calcium Ammonium Nitrate in acid soil (pH 5.5-6.5), Urea in neutral soil (pH 6.5-7.5), and Ammonium Sulphate in alkaline soil of pH above 7.5.

(*iii*) Aquatic Weeds and Insect Control

Aquatic weeds are of common occurrence in fishery waters and are generally undesirable. Besides consuming the nutrients from the water body, excessive growth of aquatic vegetation poses serious problems by upsetting the oxygen balance, creating obstruction in movement of fishes and in netting operations etc. The weeds are, thus, completely removed as a first step in fish culture. But the Jhora fishery is exception from the above. Because there is lesser occurrence of aquatic weeds, only manual methods are used to control such infestations. This is convenient and economical particularly in small ponds. Aquatic insects such as *Corixa, Ranatra, Gerris, Notoneta, Cybester, Damselfly Nymph, Dragonfly Nymph* in their larval or adult stages are highly predatory on spawn, fry or fingerlings. Otters (*Lutra lutra*, Linnaeus) and water shrew (*Neomys fodiens*, Schreber) are permanent enemies of the uplands Jhora fishery culture. More effective control of the same can be done by the application a soap and oil emulsion 24 hours prior to stocking of spawn.

3.9. Stocking Management

(a) Species Combination

Judicious selection of compatible fast growing species is of vital importance in maximizing fish production. Selection of species ratio generally depends on local conditions of seed availability, nutrient status of pond etc. A combination of species in different, *viz.* surface layer : Catla(*Catla catla*)/Silver carp (*Hypophthalmichthys molitrix*); middle layer : Rohu (*Labeo rohita*)/*Labeo dero*/Grass carp (*Ctenopharyngodon idella*); bottom layer : Mahasol (*Tor tor; Tor putitora*)/Katli (*Neolissocheilus hexagonolepis*),/Asala (*Schizothorax richardsonii*),/Mrigal (*Cirrhinus mrigala*) and Common carp (*Cyprinus carpio*) fulfills the criteria of species selection and has proven to be the ideal combination for *Jhora* fish culture in Darjeeling, West Bengal. However, not much work has been executed on *Jhora* fish culture, the stocking combination and densities are selected mostly on an ad-hoc basis. Most farmers prefer monoculture while to stocking indigenous fish species. Although the Katli are bottom feeder, polyculture with other species preferring a different niche is not really a viable option since the depth of *Jhora* ponds do not support a 'niche-segregation'. Amongst the prominent culturable indigenous coldwater fish species of Darjeeling district, the snub-nosed Chocolate Mahseer, *Neolissocheilus hexagonolepis*, also known as Katli, is most popular among Jhora farmers. Katli is considered as a most prefered species for culture because of its long spawning period, very unique and swift migratory habit and its ability to withstand stress and undergo spawning in varying environmental conditions (Plates 31–35).

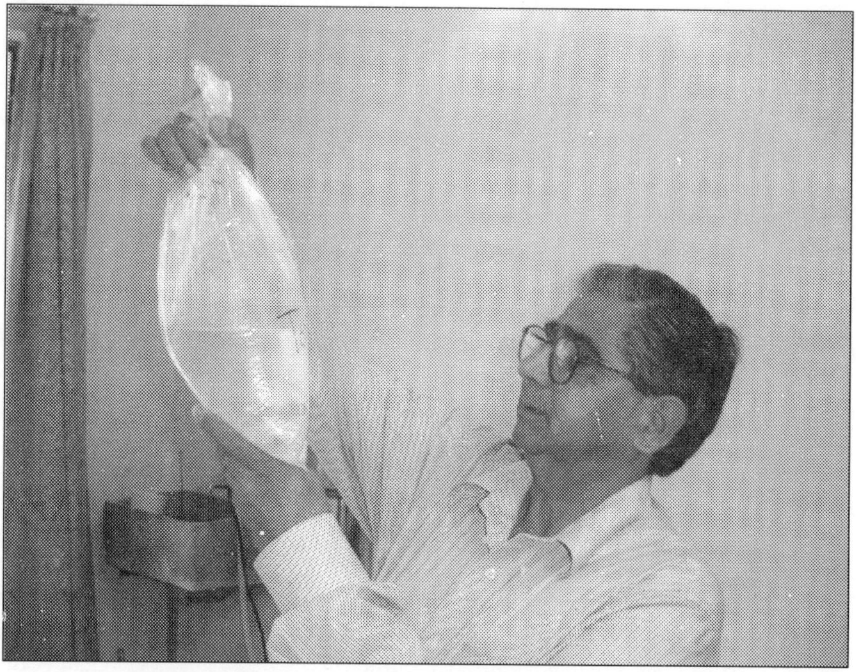

Plate 31: Health Check of Fingerlings being Done by Coldwater Fisheries Scientist at Reyang Fish Seed Farm

Plate 32: Liberation of Seed in Jhora Fish Pond by Sri Kiranmoy Nanda, Hon'ble MIC of Fisheries, at Bungklong in Mirik Block

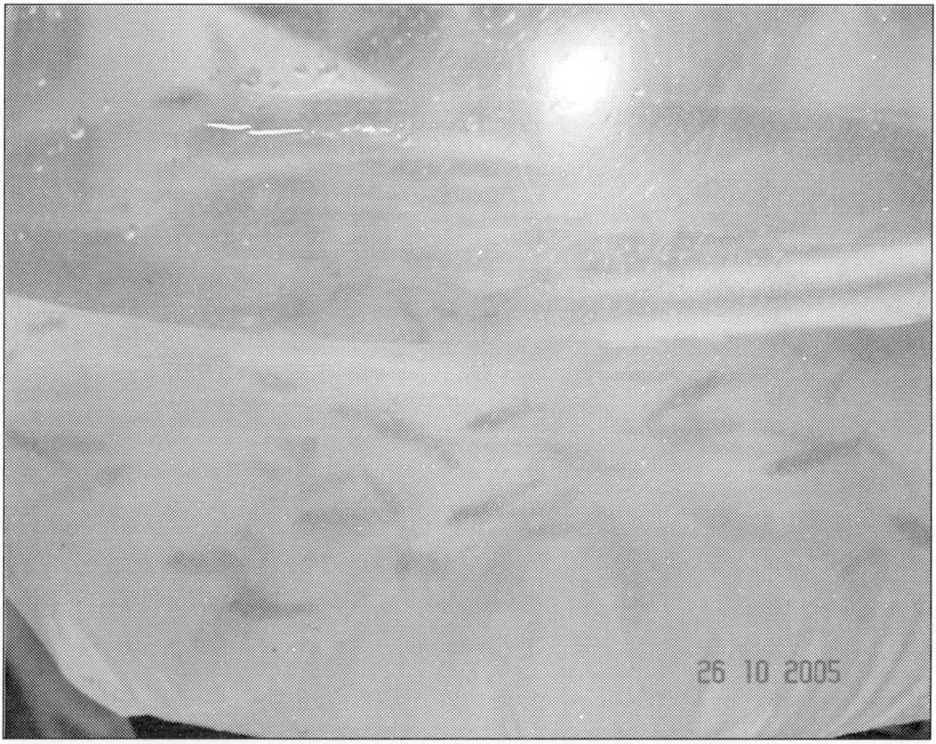

Plate 33: Oxygen Packed Fish Ready for Distribution

Plate 34

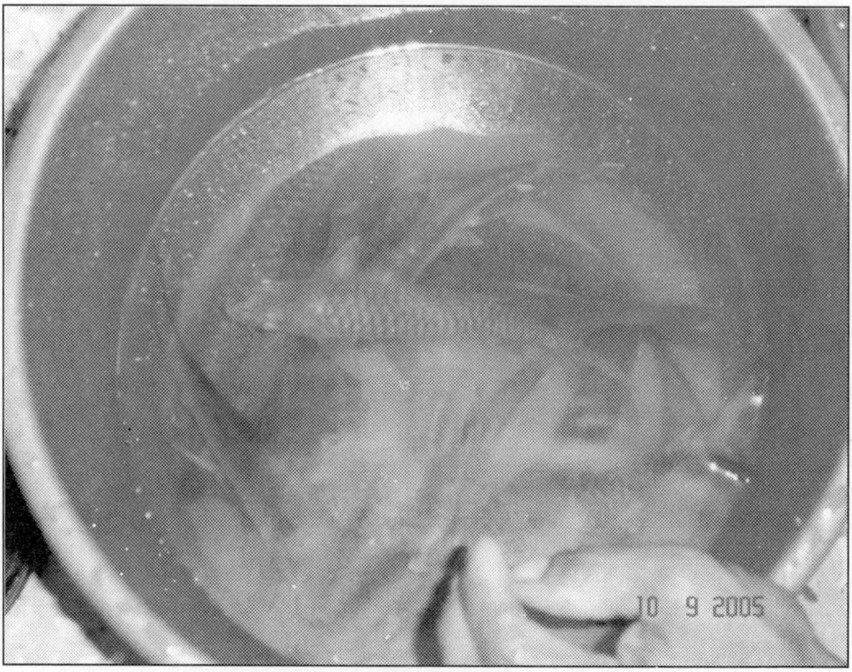

Plate 35

Plates 34–35: Fingerlings of *Tor putitora* Collected from Reyang, Rangeet and Teesta

(*b*) Stocking Rate and Size

Rate of stocking generally depends on the fertility of the pond and the measures adopted to increase its biological productivity by fertilization and artificial feeding etc. Stocking rate is generally decided in relation to area of the pond at the time of stocking. The density is considered high when stocking is done at about 15,000 to 20,000 fingerlings of about 7 – 8 cm size per ha. Fingerlings @2240 nos/unit of 1500 sq ft (140 M^2) area (*i.e.*1, 60,000 nos/ha (10000 M^2) are suitable for stocking. It is generally advisable to stock the ponds with fingerlings of 100-150 mm size for better survival. The culture period is usually ranges from March to October.

(*c*) Time of Stocking

While water temperatures ranging from 15-25°C is conducive. Thus, for better fish growth, low temperatures below 6 °C adversely affect the growth on such areas of severe winter the stocking time need to be adjusted accordingly so that period of lesser temperate (winter months) are not included as the culture period. A survey of some Jhora ponds revealed that indigenous coldwater fish species have a better growth rate, compared to Indian Major Carps. The growth of Katli has been reported to a maximum of 250 – 300 g in one season. However, growth rate varies with culture conditions. In an experimental Jhora culture in Pudung village near Kalimpong, the fish could only attain about 250 g in six months.

3.10. Transit Tanks

This is a recent concept of acclimation of fish seed with a view to reducing mortality during on transit. As fingerlings are brought from the plains and are taken to different heights, acclimatization to ambient temperature, soil water quality and other environmental situations are necessary. Fingerlings of common carp and grass carp are brought from the plain land and subsequently are acclimatized at the Hill Transit tanks (Plate 36), liberation of the same in the Jhora during July and August. Such Transit farms in the Block Headquarters facilitate distribution of fish seed to the farmers of the remote rural areas. These farms are equipped with Oxygen cylinders, polythene bags and other requisites for packing of fish seed for proper transport. Six transit tanks have been constructed by the Department of Fisheries, Government of West Bengal and Darjeeling Gorkha Hill Council (DGHC) at

(*i*) Lodhama (Pool Bazar – Bijanbari Block)

(*ii*) Lebong (Bijonbari)

(*iii*) Lamahatta (Rangli –Rangliat)

(*iv*) Ramgbhang (Jor bunglow – Sukhiapokhri)

(*v*) Kalimpong block-1

(*vi*) Majitar

Steps are being taken to construct four more such units in the places namely.

(*i*) Gorubathan,

(*ii*) Mungpoo

Plate 36: View of a Transit Tank

(*iii*) Sittong

(*iv*) Todhay Tangtta

The impact of the transit tanks has been very fruitful. One eco-hatchery has also been set up at Reyang by the State Fisheries Department with the help of DGHC to cater to the need of the farmers of the uplands areas.

3.11. Post stocking management

(*i*) Supplementary Feeding

The level of natural food organisms in the fish culture ponds can not be maintained to the required extent even after manuring and fertilization. Hence, supplementary feeding with feeds enriched with protein and carbohydrates etc. is essential for proper fish growth.

(*a*) Criteria of Food

Protein requirement of about 28-30 per cent have been found to be essential for optimum growth in carps in hills. However, to economize on the Cost of feed, a reasonably nutritive mixture of supplementary feed may contain 25-30 per cent protein, 30-40 per cent carbohydrate with dietary energy content at 3.5-4 K cal/g feed. Acceptability, digestibility, assimilation and conversion efficiencies are other criteria for selection of an efficient feed. Basic characteristics of desirable Feed are given below:

(*i*) Compact with minimum dust.

(*ii*) Stable in water, swelling and softening after absorption of water but not soluble.

(*iii*) Higher growth rate;

(*iv*) Alternative and palatable with high F.C.R value;

(*v*) Higher water quality;

(*vi*) Higher the feed price but lowers the cost of production.

(*vii*) Higher the fish health and lower the incidence and severity of disease;

(*viii*) Nutritionally balanced to promote growth.

(*b*) Feed Items

Various items of vegetable and animal origin are used as feed of carps. The vegetable items include leaves, grasses, tubers, roots, starches, oilcakes, grain fodders and other agricultural by-products, while the items of animal origin include chironomids, housefly maggots, fish meal, prawn meal, silkworm pupae etc. Commercial fish feed under different trade names are also now available in the country.

(c) Common Feed and the Schedule

The most commonly used mixture of oilcake of groundnut or mustard and rice polish or rice bran in 1:1 ratio by weight with conversion ratio of 2.5: 1 is an efficient supplementary feed for major carps (excepting for grass carps.) Other oilcakes like Til (Sesamum) and coconut can also be used. A feeding rate at 2-3 per cent of body weight of fish per day commencing from second day of stocking is generally recommended.

A monthly sampling of fishes by drag netting is essential for estimating the average weight of fishes, and thus the standing crop, for calculation of feed quantity required. However, the quantity may be regulated according to consumption etc.

A generalized feeding schedule is following by providing oilcake and rice barn mixture (1:1 ratio)at stocking density of 5000 fingerlings/ha. Fishes are daily fed during morning hours with doughs prepared by the feed mixture and placed in trays hung from a bamboo fulcrum at different depths of the pond. In case of appearance of algal bloom or oxygen depletion due to any other reason, feeding need to be suspended till revival of normal conditions.

(*d*) Feed for Grass Crap

Grass carp generally fed with halhala (local herb-*Erchities valerio folia*) or squash leaves grown in almost all houses in hill areas. Sometimes other foliage of local plants or vegetables also fed. Common carp generally feed on faecae extract of grass carp and half digested feed of grass carp. Cut bits of marginal vegetation, soft fodder grasses like Napier grass, be seem and soft vegetable wastes etc. can also be fed. Weeds may be dumped in corners of the pond or placed on an elevated platform made of coconut rope tied in bamboo frame and kept floating. The feeding schedule is maintained according to temperature, since growth of fish fed intensively is directly proportional to the extent of water temperature higher than 14°C.

The feeding schedule as practiced according to temperature:

15°C–18°C; 2 per cent Body weight

18°C – 21°C; 3 per cent Body weight

21°C – 24°C; 5 per cent Body weight

Above 24°C; 5 per cent Body weight

However, at high density like intensive culture in running water Jhora fisheries of Darjeeling hills, fish are dependent on natural food and some quantity of locally available supplementary feed. Thus in intensive fish culture inadequacy of supplementary feed lead to poor growth, nutritional disease and in poor fish condition, increased susceptibility to parasite and bacterial infection.

(*ii*) Growth Performance of Fish Species

Common carp and grass carp fingerlings when stocked in the *Jhora* grow upto 500 gms to 800 gms (average being 300-400 gms) in one year period. The culture period for Indian Major Carps varies between 6 to 9 months (March to November), whereas exotic carps are cultured throughout the year. Observation on the carp culture in the hills (altitude 4000ft) show that Rohu and Mrigal can be cultured suitably in some selected pockets of hills (lower altitude *viz.* Reyang etc.). Catla cannot survive without stress in coldwater ecological conditions. With a culture period of 3 years Rohu is reported to be grown to a size of 40 cms,–1010 gms in weight and Mrigal to a size of 45 cm–1240 gms. Common carp can survive in existing ecological suitably and breeding was observed in lower altitude (alt. 2500 ft.). In Darjeeling Hill areas temperature varies from 0° C to 20° C and thus growth of fishes is observed to be slow in comparison to the plain land. However, it has been observed harvesting fish at 400-500 gms size is more economic and give higher yield. Mortality rate is on and average 20 per cent, if management is proper and fingerlings stocked after due acclimatization. The trend on growth of fish is depicted below:

Species	*Growth Trend and Culture Period*
Catla (*Catla catla*)	: 250-350 g in six to nine months
Silver carp (*Hypophthalmichthys molitrix*)	: 550-950 g in twelve months
Rohu (*Labeo rohita*)	: 350-400 g in six to nine months
Kursa bata (*Labeo dero*)	: 150-200 g in twelve months
Grass carp (*Ctenopharyngodon idella*)	: 950-1050 g in twelve months
Mahasol (*Tor tor; Tor putitora*)	: 250-500 g in twelve months
Katli (*Neolissocheilus hexagonolepis*)	: 550 -650 g in twelve months
Asala (*Schizothorax richardsonii*)	: 200-250 g in twelve months
Mrigal (*Cirrhinus mrigala*)	: 250-300 g in six to nine months
Common carp (*Cyprinus carpio*)	: 6000-1000 g in twelve months
Gold fish (*Carassius auratus*)	: 150-200 g in twelve months

(*iii*) Periodical Monitoring of Ponds

This generally involves (*a*) Maintaining a constant water depth and volume and standard water quality conducive of fish production in these ponds (*b*) Removal of feed remnants and grasses and (*c*) Prevention of enemies like, otters and snakes. In *Jhora* ponds water quality found to be optimum for better fish production is summarized in Table 48.

The plankton diversity in the ponds did not seem to be high. Among the zoo-plankton, Daphnis and Cyclops were observed occasionally. Spirogyra, Chlorella, Synedra, Phacus and Navicula were observed among the phytoplankton.

Table 48: Optimum Water Quality Parameters of Jhora Ponds

Sl.No.	*Parameters*	*Range*
1.	Water temperature (°C)	8.0–28.0
2.	Water pH	6.5–7.3
3.	Dissolved Oxygen (mg^{-lt})	4.2–8.4
4.	Free CO_2 (mg^{-lt})	0.8–6.3
5.	Total alkalinity (mg^{-lt})	18.5–35.4
6.	Total hardness (mg^{-lt})	9.6–26.8
7.	Chloride (mg^{-lt})	13.4–29.8
8.	Phosphate–P (mg^{-lt})	0.02–0.09
9.	Nitrate –N (mg^{-lt})	0.03–0.16
10.	Nitrite – N (mg^{-lt})	Traces
11.	Ammonium – N (mg^{-lt})	0.007–0.056

3.12. Harvesting

Due to retarded growth of fishes during the winter period, (November–February), the fish are harvested after a culture period of 8 months, that is, at the end of October. A brief report on harvesting (Plates 37–52) is depicted in Table 50. The harvesting process is very easy, quick and simple. At first, the inlet pipe from where the water is drawn into the depression is closed and outlets are let open. The whole process is over within an hour. The fish are hand-picked and collected in bamboo netted baskets. Since, Jhora units are small and managed by a single farmer/family, individual farmers tend to alter the fishery management according to their convenience. Many a times, fishes are not cultured for 8 months culture period. Instead, they partially harvest (phased harvesting) after 4 – 5 months of culture and new fingerlings are stocked along with the remaining fishes of the earlier stock. Average yield per unit is 350-400 kg/year. During such harvesting, fish are netted out instead of withdrawing water from the pond. The fish are locally sold at prices ranging from Rs.60 – 100 per Kg. Harvesting devices are as follows:

(*a*) A major feature of the conventional cast net used in uplands Jhora fishery is that solid iron or lead sinkers, weighing about 5.0 kg are provided on the peripheral cord. Cast net is employed to catch minor carps as well as major caps and other fishes.

(*b*) Drag net is used for community fishing in conjunction with stake net and cast net.

(*c*) Single head spears, multiple-head spears (Narsoo) and long lines are also used in many areas.

Plates 37–38: Catching of Wild Fish at River Teesta Near Sevoke Using Cast Net

Plate 37

Plate 38

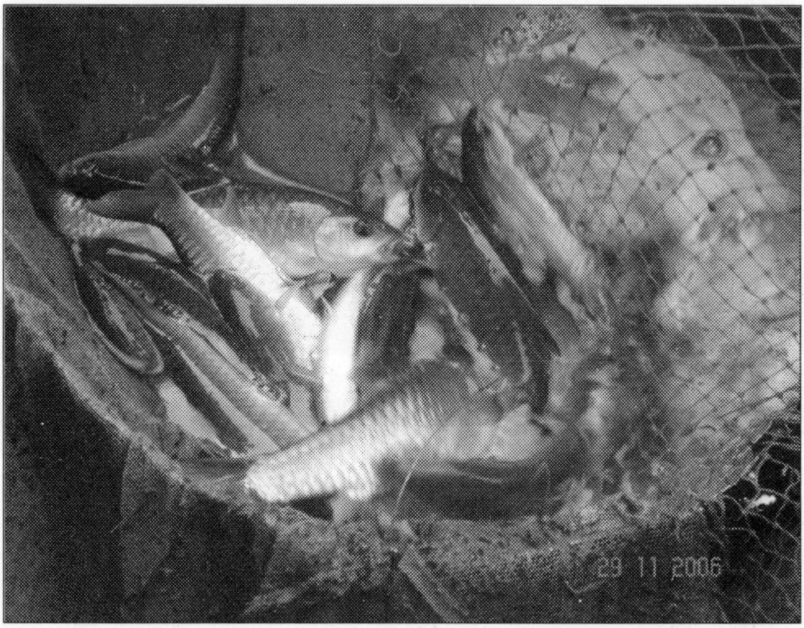

Plate 39: Harvest of *N. hexagonolepis* from Jhora Fish Pond

Plate 40: Wild Catch of Snow Trout

Plates 41–42: Harvest of Composite Fish Culture from Jhora Fish Pond

Plate 41

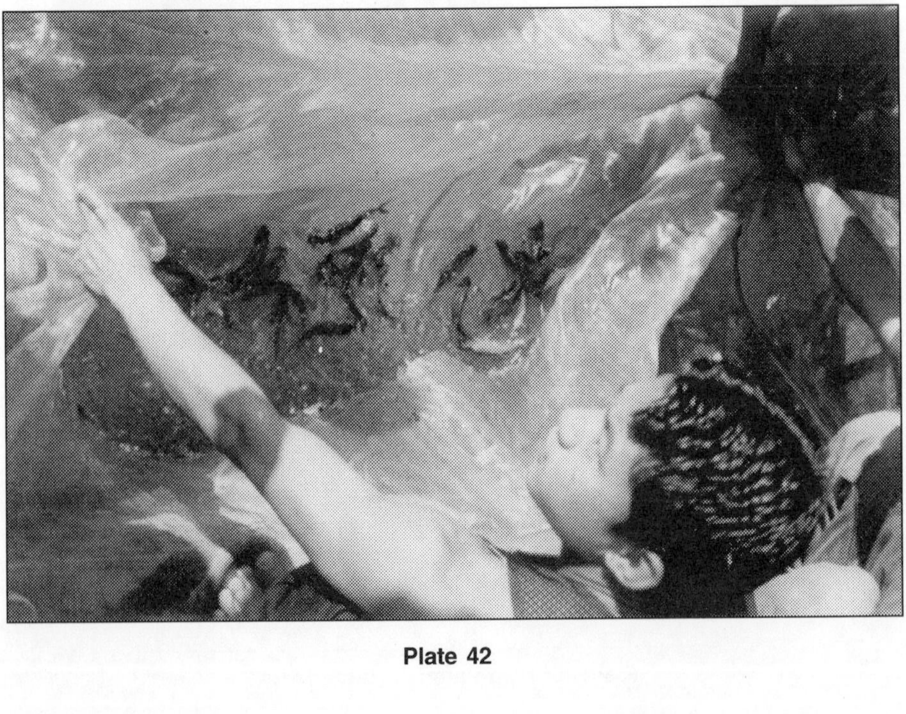

Plate 42

Plates 43–44: Collection of Coldwater Fish Species from the River Teesta

Plate 43

Plate 44

Plates 45–48: Different Types of Traditional Fishing Gears Used in Hill Stream Rivers

Plate 45

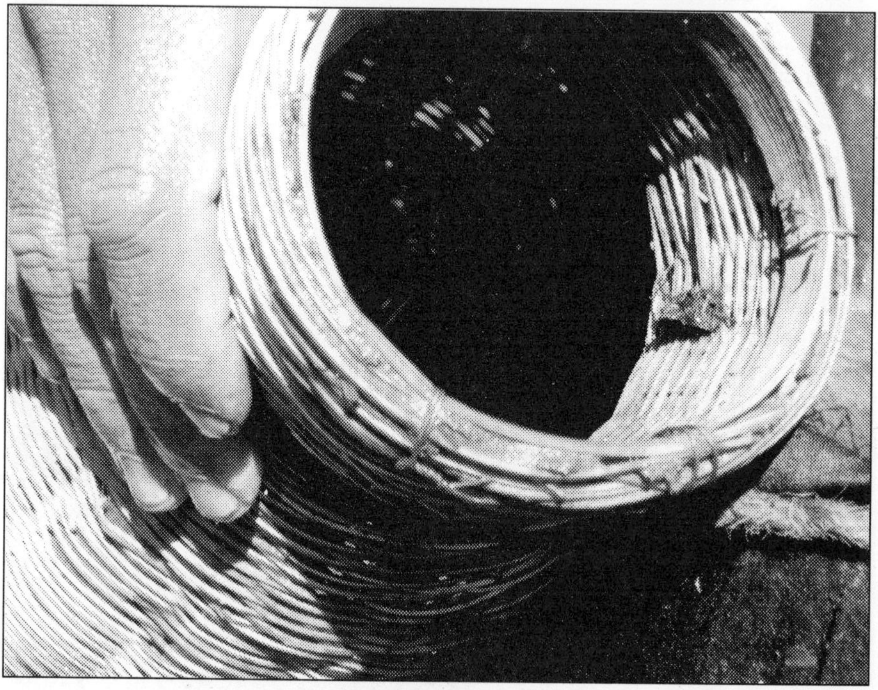

Plate 46

Plate 47

Plate 48

Plates 49–52: Harvesting of Coldwater Fishes

Plate 49

Plate 50

Plate 51

Plate 52

3.13. Marketing

After harvesting the producer sale fishes directly in the retail market (Plates 53–60) for fresh consumption of the fishes.

Channels

Producer ⟶ Retailer ⟶ Consumer

Producer ⟶ Consumer

Plates 53–58: View of Various Local Species Sold at Different Retail Market in North Bengal

Plate 53

Plate 54

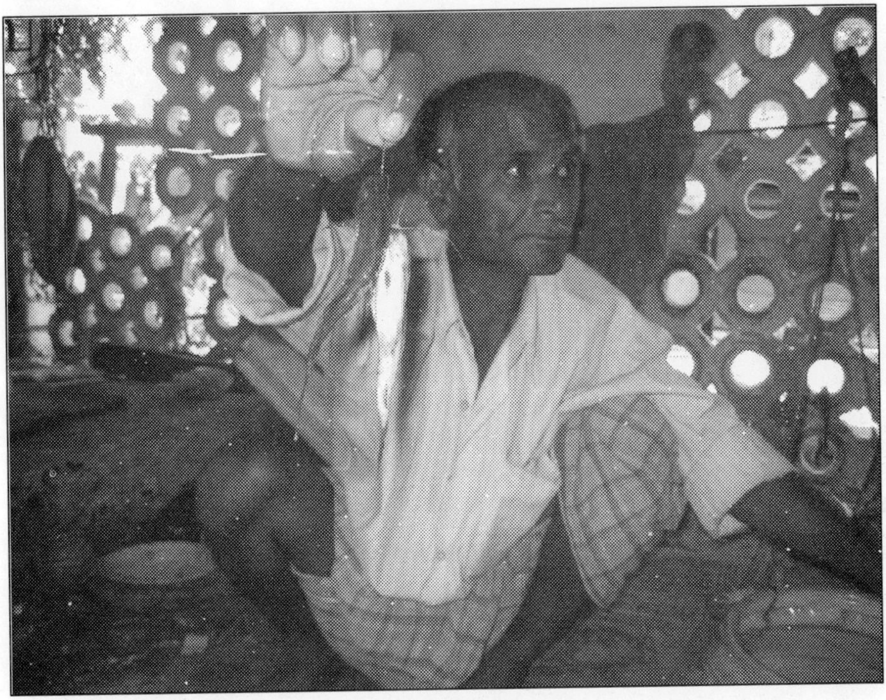

Plate 55

Plate 56

Plate 57

Plate 58

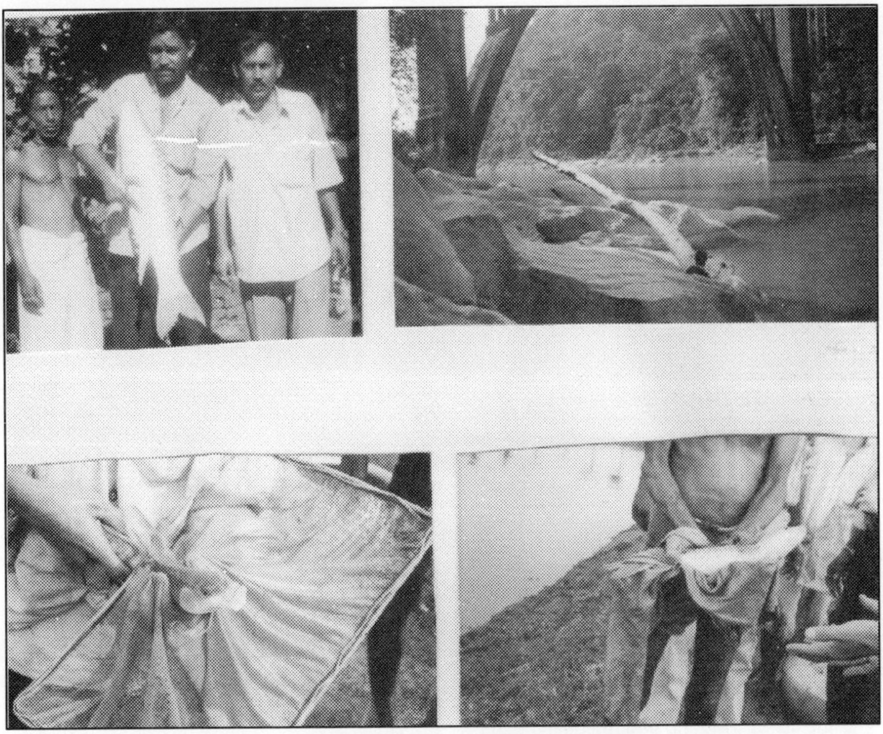

Plate 59: Prize Catch of *Tor putitora* from River Teesta

Plate 60: Indigenous Gear Used for Catching Fingerlings from Hill Streams

3.14. Economics

It is most essential to know before implementing any such development programmed to consider their economic viability. The hilly areas are mostly far behind in development and still in many areas agriculture development are limited due to topography of the area. The higher fish yield achieved will reduce the cost of yield and the programmes may be economically viable. Supplementary source of earning and employment potentialities, these programmes will be beneficial. Apart from these advantages, it will also help to provide revenue to the state.

Pond fisheries fed by Jhora water in the hills of Darjeeling (Unit size 1000 square feet (25ftx40 ft) with 1.2 mt depth)

A. CAPITAL COST

1. Earthwork

		Quantity	Rate	Amount (Rs.)
(i)	Earth work in excavation from borrow pits Depositing inlayers of 250 mm including breaking of clods in mix hard soil with an initial lead of 50 m and initial lead of 1.5m making bunds at top as directed and dressing complete	60 cum	36	2160
(ii)	Turfing with grass sods 5 cm to 8 cm thick including watering and maintaining in fully grown	23 sqm	6.4	147.2
(iii)	Supplying and laying stone boulder stone set edging with ruff dressing the boulder cutting necessary trenches in all class of soil and sidep acking with earthen bunds including cost of materials and labour all complete.	23 sqm	360	8280
				Total Rs.10587
				Say Rs.10600
2.	Rubber tubing of dia 1" (1 kg: 10 feet)	50 kg	35	1750
				Sub Total Rs.12350
B.	**RECURRING COST**			
(i)	Lime	9 kg	5.70	51
(ii)	Fingerling (7-10 cm size) (Stocking @16 nos/M²)	2240	1.30	2912
(iii)	Fish Feed (FCR 1:1.5)	675 kg	12.00	8100
(iv)	Netting, Prophylactics, collection of grass etc.	L.S.	200	
				Sub Total 11263
		Grand Total (1+2)		23613
		Say		23600

	Quantity	Rate	Amount (Rs.)

C. Unit Cost **Rs. 23600**

D. Margin 15 per cent **Rs. 3540**

E. Bank Loan **Rs. 20060**

F. Income **Rs. 19250**

Fish production 350 kg @Rs. 55/–and survivability 60 per cent, Average wt. at harvest 350 gm.

Note: Subsidy available in case of FFDA beneficiary @20 per cent of the project cost or max of Rs.5000/. Ceiling of 3 units for each farmers in terms of admissibly.

G. FINANCIAL ANALYSIS

(*i*) NPW AT 15 per cent DF 2759

(*ii*) BCR AT 15 per cent DF 1.05:1

(*iii*) IRR 20.55 per cent

H. REPAYMENT SCHEDULE

Bank loan	20060	2407	22467	7700	2.92
Repayment schedule	Four years with first years as grace period				
Repayment Insttalment (12 per cent) interest		I year	I lyear	III year	I Vyear
Annual equited installments–4 years		6604	6604	6604	6604
Capital recovery factor-0.0329234					
Net surplus (after payment)		–18941	1382	1382	1382

I. Income expenditure statement and Repayment

	Particulars	Years			
		I	*II*	*III*	*IV*
1.	Gross Income from fish	0	19250	19250	19250
2.	Gross expenditure	23600	11263	11263	11263
3.	Gross Surplus (1-2)	–23600	7987	7987	7987
4.	Add recurring cost for term loan	11263	0	0	0
5.	Amount available for the repayment (3+4)	–12337	7987	7987	7987
6.	Less AEI (12 per cent) CRF:0.329234	6604	6604	6604	6604
7.	Net surplus (5-6)	–18941	1382	1382	1382

J. FINANCIAL ANALYSIS

NPW 2759

BCR 1.13:1

IRR 20.55 per cent

Year	1	2	3	4	5	6
Cost	23600	11263	11263	11263	11263	11263
Benefit	0	19250	19250	19250	19250	19250
Net Benefit	–23600	7987	7987	7987	7987	7987
PWC	53353					
PWB	56112					
IRR	20.55 per cent					

3.15. Ongoing Schemes in the Hill Areas of Darjeeling District

Darjeeling Gorkha Hill Council (DGHC) in coassociation with the Department of Fisheries, out of Government of West Bengal has taken up many schemes since its inception which may broadly be classified as follows:

1. Normal Scheme
2. Fish Farmer Development Agency (FFDA) Scheme

1) Under the normal scheme we have individual as well as community schemes

(*a*) Under the individual scheme we have schemes like

I. Construction of individual ponds for the people living in model villages
II. Providing subsidy to individual farmer
III. Distribution of mini kits
IV. Distribution of fingerlings free of cost
V. Schemes under Tribal Sub Plan
VI. Training programme
VII. Construction of houses for individual farmers living in model village

(*b*) Under the community scheme we have schemes like

I. Construction of transit ponds
II. Construction of rearing ponds
III. Construction of aquarium etc.

To meet the requirement of healthy fast growing and disease free fingerlings of our farmers to some extent a Fish Seed Farm has been set up at Reyang, under Kalimpong I Block on N.H. 31, which was commissioned for operation on 23rd April 2002 by Hon'ble Minister-in-charge, Fisheries Department, Govt. of West Bengal. This is the first fish seed production farm setup in the hill areas of Darjeeling. The farm is also used to impart on farm training facilities to the JHORA beneficiaries. In future the Department intends to construct similar type of farms in various areas of Darjeeling Hills so that our fish farmers may not have to face any difficulties in producing fesh seeds from such farms as and when required.

During the year 2001-2002, a new scheme called "Fishermen Model Village" to boost up Fishery activities in the hill areas of Darjeeling were taken up by the Department. A village called "BUNKULUNG" under Mirik Development Block, about 20 km from Mirik was taken up as Model Village (Plate 61). Under the Scheme a total number of 20 houses and 17 numbers of individual ponds were constructed for the benefit of the farmers living in the said village. In order to preserve their culture and also to provide them entertainment a community hall has also been constructed.

To educate the people on conservation of hill fishes and also to boost up tourism in the hill areas of Darjeeling District the Department has taken up a project of construction of an aquarium at Mirik which is under construction. Besides for sport loving people, a proposal for construction of angling platform around Mirik lake is also in hand which in near future will be a source of revenue to the Department.

Plate 61: Model Fishermen Hut Under Construction at Bungklong in Mirik Block

In order to conserve the hill fishes which thrive in coldwater and also to increase their production for propagation a proposal to set up a "Coldwater Fish Farm" at Relli Pala under Kalimpong II Development Block is also in hand and the proposal has already been sent to the Director of Fisheries, Government of West Bengal for further process. The work on the said project will start as soon as the necessary fund is available from the Government.

With the introduction of all these schemes in the hill areas of Darjeeling for the benefit of our farmers (Plates 62–63), a momentum has already been gathered for adopting fish farming as a supplementary source of income to improve their livelihood.

Plates 62–63: Jhora Beneficiaries in Darjeeling District

Plate 62

Plate 63

2) Under the F.F.D.A scheme there are four different schemes, namely the "Jhora Fishery Scheme" and three "short term credit loan" schemes. The total project cost of Jhora fishery scheme is Rs. 27,400/–having 40 per cent subsidy for the general beneficiaries (20 per cent from F.F.D.A and 20 per cent from D.G.H.C) and 50 per cent subsidy for S.C/S.T beneficiaries (25 per cent from F.F.D.A and 25 per cent from D.G.H.C).

3.16. Short-term Credit Loan Schemes Under Jhora Fishery

A Jhora Unit is a bankable scheme-enjoying subsidy under FFDA. The general scale of subsidy is 40 per cent and for SC and ST 50 per cent of which the Government and DGHC contribution is 50:50. Under approval from NABARD the project cost unit is Rs. 21,800 of which 60 – 50 per cent is bank financed. This is well within the scale for Centrally Sponsored Schemes. There are 3 other STCP schemes without subsidy:

(*i*) Culturable scheme costing Rs. 10,800/-.

(*ii*) Integrated Pisciculture with mushroom. Scale Rs. 11,900/-

(*iii*) Integrated Pisciculture with piggery. Scale Rs.15,350/-.

The "short term credit loan schemes" is further subdivided into three hands namely:

a) Culturable scheme the total project cost for which is Rs. 10,800/–with no subsidy per unit area (1 unit=1500sq ft or 140 sqm)

	Short Term Credit Loan	*Amount (Rs.)*
A.	CAPITAL COST	
1.	Minor repairing and renovation of existing JHOR tank including repairing of water intake system (one time only)	5000.00
B.	CULTURE COST	
(*i*)	Lime 9 kg	56.00
(*ii*)	Fingerlings of standard size (10cm) 2240 no. @Rs. 600.00/thousand	1344.00
(*iii*)	Fish Feed 600 kgs @ Rs. 7 per kg	4200.00
(*iv*)	Periodical Netting etc.	200.00
	B–Total	5800.00
	A + B (Rs. 5000.00+ Rs. 5800.00) (Rupees Ten thousand eight hundred only)	10,800.00
C.	PRODUCTION and INCOME	
(*i*)	Selling of fish 400kg @ Rs. 40.00 per kg	16,000.00
D.	SOURCE OF FUND	
	Bank Loan 100 per cent	10,800.00

	Short Term Credit Loan	*Amount (Rs.)*
E.	ECONOMICS	
(*i*)	Gross income by fish	16,000.00
(*ii*)	Repayment of Bank loan (int as per Bank norms)	10,800.00
(*iii*)	NET PROFIT (less Bank int.)	5200.00

Note: Capital cost as shown in A is the one time expenditure and hence from the 2nd year beneficiary will earn more profit from the above scheme. Insurance coverage should be made.

b)Integrated pisciculture with mushroom the total project cost for which is Rs. 11,900/- per unit area (1 unit=1500sq ft or 140 sqm) with no subsidy

	Integrated Pisciculture With Mushroom	*Amount (Rs.)*
A.	CAPITAL COST	
1.	Minor repairing and renovation of existing JHORA tank including repairing of water intake system (one time only)	5000.00
B.	CULTURE COST	
(*i*)	Lime 9 kg	56.00
(*ii*)	Fingerlings of standard size (10cm) 2240 no. @Rs. 600.00/thousand	1344.00
iii)	Fish Feed 600 kgs @ Rs. 7 per kg	4200.00
(*iv*)	Periodical Netting etc.	200.00
(*v*)	Mushroom culture (seed, straw, structure etc.)	2500.00
	B–Total	6900.00
	A + B (Rs. 5000.00+ Rs. 6900.00) (Rupees Eleven thousand nine hundred only)	Rs. 11900.00
C.	PRODUCTION and INCOME	
(*i*)	Selling of fish 400kg @ Rs. 40.00 per kg	16000.00
(*ii*)	Selling of mushroom	5000.00
	Total Income	21000.00
D.	SOURCE OF FUND	
	Bank Loan 100 per cent	11900.00
E.	ECONOMICS	
(*i*)	Gross income by fish and mushroom	21000.00
(*ii*)	Repayment of Bank loan (int as per Bank norms)	11900.00
(*iii*)	NET PROFIT (less Bank int.)	9100.00

Note: Capital cost as shown in A is the one time expenditure and hence from the 2nd year beneficiary will earn more profit from the above scheme. Insurance coverage should be made.

c) Integrated pisciculture with pigs the total project cost for which is Rs. 15,350/- per unit area (1 unit=1500sq ft or 140 sqm) with no subsidy

	Integrated Pisciculture With Pigs	*Amount (Rs.)*
A.	CAPITAL COST	
1.	Minor repairing and renovation of existing JHORA tank including repairing of water intake system (one time only)	5000.00
B.	CULTURE COST	
(*i*)	Lime 9 kg	56.00
(*ii*)	Fingerlings of standard size (10cm) 2240 no. @Rs. 600.00/thousand	1344.00
(*iii*)	Fish Feed 300 kgs @ Rs. 7 per kg	2100.00
(*iv*)	Periodical Netting etc.	200.00
(*v*)	Piglets 45 days old exotic variety 2 female and 1 male @ Rs. 550.00 per no.	1650.00
(*vi*)	Pig feed	4000.00
(*vii*)	Pig Sty including Medicine and Insurance	1000.00
	B–Total	10350.00
	A + B (Rs. 5000.00+ Rs. 10350.00) (Rupees Fifteen thousand three hundred fifty only)	Rs. 15350.00
C.	PRODUCTION and INCOME	
(*i*)	Selling of fish 400kg @ Rs. 40.00 per kg	16000.00
(*ii*)	Selling of pigs 250kg @ Rs. 40.00 per kg	10000.00
C.	Total Income	26000.00
D.	SOURCE OF FUND	
	Bank Loan 100 per cent	15350.00
E.	ECONOMICS	
(*i*)	Gross income by fish and pig	26000.00
(*ii*)	Repayment of Bank loan (int as per Bank norms)	15350.00
(*iii*)	NET PROFIT (less Bank int.)	10650.00

Note: Capital cost as shown in A is the one time expenditure and hence from the 2nd year beneficiary will earn more profit from the above scheme. Insurance coverage should be made.

3.17. Advocation of Jhora Fishery in the Hills of Darjeeling District

Before 1980, pisciculture in the upland of Darjeeling was a doubt. But after observing the rate of growth of fishes over the years in artificial impoundments, fed by perennial Jhora, the Govt. felt encouraged to advocate Jhora water fisheries in the hills to solve the acute protein need in the remote hill areas. When scientifically cultured Jhora Fisheries have the additional advantage of

(*i*) being virtually maintainable as free from pollution
(*ii*) catch ensured due to confinement
(*iii*) good survival for indigenous species
(*iv*) low cost
(*v*) sustainable
(*vi*) suitable for rain water harvesting
(*vii*) ecologically sound

Table 49: Block-wise Distribution of FFDA, Darjeeling Subsidy Released from 1990–2006

Sl.No.	*Year*	*A**	*B*	*C*	*D*	*E*	*F*	*G*	*H*	*I*
1.	1990-1991	549	92	64	54	107.5	61	405	108.5	21.5
2.	1991-1992	371	49	52	12	90.5	15	29	92.5	31
3.	1992-1993	90	24	06	10	39	53	04	01	01
4.	1993-1994	197.5	06	11	34	66	31	11	24.5	14
5.	1994-1995	188	17	02	15	61.5	41	16	23	12.5
6.	1995-1996	169	16	13	08	57	23	25	22	03
7.	1996-1997	149	02	38	00	55	16	04	10	24
8.	1997-1998	115	17	04	16	26	20	00	32	00
9.	1998-1999	150	23	09.5	15	34	17	05	14	32.5
10.	1999-2000	142.5	27	10	13	29	27	13	05	18.5
11.	2000-2001	119	35	29	.00	27	13	11	00	04
12.	2001-2002	171	58	23	34	27	18	00	01	10
13.	2002-2003	143	46	38	00	33	01	00	09	16
14.	2003-2004	122	08	14	44	21	29	00	00	06
15.	2004-2005	154	21	09	48	41	05	00	10	20
16.	2005-2006	118	32	17	16	05	33	00	00	15
	Total	2948	473	339.5	319	720.5	403	158.5	352.5	229

* A: Sanctioned Units; B: Darjeeling-Pulbazar; C: Jorbunglow-Sukhiapokhari; D: Rangli-Ranliot; E: Kalimpong-I; F: Kalimpong-II; G: Gorubathan; H: Kurseong; I: Mirik.

3.18. Hazards of *Jhora* Fishery

(*i*) Natural Calamities

(a) Landslide

Landslide occurrence is very common in the hilly tracks especially during monsoon. Landslide is the result of a wide variety of processes which include geological, geomorphological and meteorological factors. The important terrain factors are lithology, structure, drainage, slope, landuse, geomorphology and road network. A complete landslide hazard assessment requires an analysis of all these factors leading to instability in the region. The feature extraction of some of these factors can be done from the interpretation of satellite images. Landslide in relation to *Jhora* fishery in Darjeeling Himalaya has been studied to decipher the influence of these parameters on landslide occurrence. Drainage density is defined as the ratio of sum of the drainage lengths in the cell and the area of the corresponding cell.

(*b*) Otter

Otters are highly pisicivorous. Especially in hill areas, due to low depth of water bodies fishes are readily available for otter. It is considered as one of the greatest enemies for the culture fisheries. A very funny indigenous method for controlling these otters are by mimicking lime filled egg shell as pure egg for luriaging the otters.Otters generally die immediately while engulfing the whole egg-shell filled with lime. Limes are use in egg-shell as safety measures for fish culture to avoid any kind of poising in the water.

(*c*) Dryness of *Jhora*

It is often observed that due to changes of topography, climate\meteorological fluctuations, the water of the *Jhora* sometimes almost reaches to dry condition resulting in no production from those water bodies.

(*d*) Change of Flow of *Jhora*

Due to obstruction in the topography, fluctuation of Ice melting and meteorological changes, the spring or *Jhora* sometimes changes its normal course from the original path. Thus, the *Jhora* practices are hampered owing to the changes in water flows.

(*e*) Deforestation and Devastating Flash Floods

Deforestation takes place due to the permanent destruction of indigenous forests and woodlands. Deforestation has resulted in the reduction of indigenous forests to four-fifths of their pre-agricultural area. It causes soil erosion, landslide, and siltation in the stream beds facilitating the devastating of flash floods in hill regions of Darjeeling district.

(*f*) Temperature

The growth and survivability of fishes depends largely on the temperature exists in its habitat. The temperature of Darjeeling hill areas varies significantly from 0–30°C. The graph of such temperature variation (Figure 10) of the area is depicted below. The fish production from the Jhoras is affected with such fluctuations of temperature.

(*ii*) Common Diseases

Uplands fish's shows the symptoms of various diseases can be categorized under three heading *viz.* behavioral, external physical and internal physical. Some of the important behavioral characteristics of diseased fish are:

(*a*) loss of appetite

(*b*) abnormally congregating in pond, like remaining on the surface,

(*c*) gathering on one side, crowding near the head or tail screen,

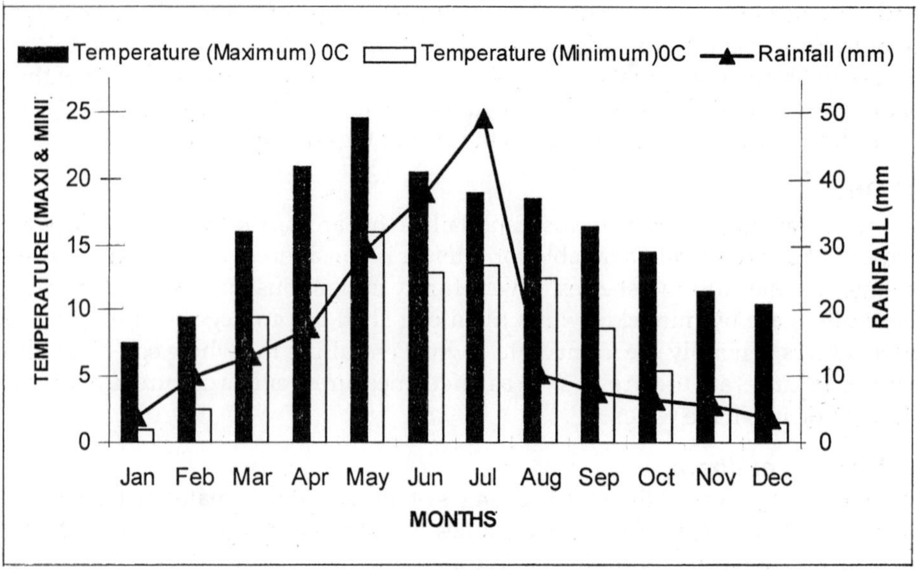

Figure 10: Monthly Variation of Temperature and Rainfall

(*d*) flashing, scraping on bottom or any other object, darting, whirling, twisting, loss of equilibrium;

(*e*) loss of vitality, weakness,

(*f*) loss of ability to withstand handling during size-grading, seining, loading or transportation.

A very wide variety of diseases are shown in uplands *Jhora* fish culture.

(*a*) Viral infections as Infectious Pancreatic Necrosis (IPN) and VHS (Viral Hemorrhagic septicemia)

The treatment of infected fishes includes dipping them in potassium permanganate of concentration 1 g/litre of water for 60-90 minutes.

(*b*) Bacterial diseases as Vibrio diseases, *Furunculosis, Columnaris* and Fin-rot

The treatment of infected fishes includes dipping them in potassium permanganate of concentration 1 g/litre of water for 60-90 minutes.

(*c*) Fungal infection

The treatment of infected fishes includes dipping them in potassium permanganate of concentration 1 g/litre of water for 60-90 minutes.

(*d*) Diseases by parasites as *Costiasis, Whirling disease, Ichtyophthiriasis or Ich disease*

The treatment of infected fishes includes dipping them in quinine solution of concentration 1 g/20 litre of water or trypoflavin solution 1 g/100 litre of water or chlororamine solution 1 g/100 litre of water have been recommended.

(*e*) Nutritional diseases

(*f*) Ulcerative disease syndrome

(*g*) Disease from uncertain origin like permanent enemies.

"Prevention is better than cure" – but some time it was happen that different types of abnormalities occur in the Jhora fish culture ponds.

(*iii*) Water Scarcity and Watershed Management

Darjeeling being a hill station suffers from water scarcity in respect of drinking, irrigation, fishery and domestic purpose. To overcome the situation only solution is to conserve water through water shed management. Although Department of Fisheries and DGHC have put thrust on watershed management the hill area but it could not suffice the demand. Physico-chemical and biological characteristics of three types of streams (Fast, medium and slow) in the upland of Darjeeling is required to be studied. Very little work on the same have, however been made. The National Watershed Development Project for Rainfed Areas (NWDPRA) was launched during 1990-91 in 25 States and two Union Territories. During the IXth Plan it was proposed to treat an area of 2.25 million hectares (in the Xth Plan it is 12 million ha) at an estimated cost of Rs.1030 crores. In 1999-2000 for the first time a Watershed Development Fund (WDF) was established at NABARD to enable States to access credit for treatment of larger areas under watershed development. The objectives of NWDPRA are as follows:

1. Conservation, development and sustainable management of natural resources including their use.
2. Enhancement of agricultural productivity and production in a sustainable manner.
3. Restoration of ecological balance in the degraded and fragile rainfed ecosystems by greening these areas through appropriate mix of trees, shrubs and grasses.
4. Reduction in regional disparity between irrigated and rainfed areas.
5. Creation of sustained employment opportunities for the rural community including the landless.

(*iv*) Other Constraints of the Coldwater Fisheries

1. Fish culture in hill area is a hobby and a secondary source of income in this region.
2. Faulty construction of the fish ponds and Pond size is less than the optimum one.
3. Lack of in farm training facilities to the entrepreneurs.
4. Unavailability of the fingerlings and other required inputs.
5. Lack of integrated fish farming approaches.
6. Lack of proper market facility for the locally produced fish.
7. Lack of regular water supply.
8. Bad Communication system.

9. Dams and barrages constructed on the river Teesta and Human settlements in Teesta and its tributaries hamper the migrations fishes.
10. Lack of awareness program among the local people for conservation, propagation and for safe guarding the local fish species.
11. Limited scope for the exploitation of local indigenous ornamental fish species.
12. Rampant destruction of natural fish habitat and breeding grounds by the local people by using unscientific fishing methods like dynamiting, poisoning, hammer fishing, electro fishing etc.
13. Unavailability of Financial Aid and lack of interest of financial institutions.

3.19. Women Involvement in Jhora Fishery

A large number of poor downtrodden women (Plate 64) folk are engaged in Jhora fisheries and sustain their families with their income. In hill region women's involvement in Jhora fish culture, integrated with animal husbandry is predominately about 80 per cent, starting from seed collection to marketing. From ancient time traditionally they cultivate exotic carps and made an important contribution to the rural economy of the area. Women of hill areas are supported by various government schemes, *viz.* training, financial support, supply of seed and feed at subsidized rates, as a results the improvement of fish production from these Jhoras.

Plate 64: Involvement of Fisher Women in Jhora Fishery

Table 50: Report on Production of Coldwater Fishes at Darjeeling as on April, 07

Sl. No.	Name and Address of the Fish Farmers	Date of Stocking	Species Found	Avg. Weight (gms.)	Avg. Length (cms.)	Survival Rate (per cent)	Production in Kg. per Unit (i.e.1500 Sq. ft.)	Remarks
1.	Shakila Lepcha, Sindebong	April'06	Asala Katli Mahsheer	250	16	52	161.6	Fish fed with food and Kitchen waste.
2.	Jong Lepcha, Sindebong	April'06	Katli	230	15	57	167	Fish fed with Kitchen waste.
3.	Gyan Lepcha, Sindebong	April'06	AsalaKatli	225	15	60	140	Fish fed with chicken guts and Kitchen waste.
4.	Dawa Lepcha, Sindebong	April'06	Katli	250	16	40	131	–do–
5.	Diwas Tamang, Sindebong	April'06	AsalaKatli	180	13	46	153.6	–do–
6.	Dip Tsh. Lepch, Sindebong	April'06	AsalaKatliv	225	12	58	173	–do–
7.	Norbu Lepch, Sindebong	April'06	Katli	240	14	59	154.1	–do–
8.	Ugen Lepch, Sindebong	April'06	Katli	190	13	41	127	Fed with variety of Vet+Non Veg feed.
9.	H.R. Pradhan, Dungra Busty	April'06	AsalaKatli	160	11	58	149.9	Good size Asala cemented tank
10.	Dinesh Mani Pradhan, Sindebong	April'06	Katli	275	15	47	161	–do–
11.	Sanchaman Lepcha, Pemling	April'06	Katli	230	13	50	145	Fed with Kitchen waste
12.	Kutbir Lepcha, Pemling	April'06	Katli	180	16	43	137	Fed with feed
13.	Mangal Lepcha, Pemling	April'06	Asala Katli Mahasheer	200	16	63	176	Fed with chicken guts+ feed
14.	La Tsh Lepcha, Pemling	April'06	Katli	200	14	44	151	Fish fed with chicken guts+ variety of herbs
15.	Yusuf Lepcha, Pemling	April'06	Katli Mahasheer	225	15	49	158.3	–do–
16.	Noramdi Lepcha, Pemling	April'06	Katli	190	14	41	133.2	

Contd...

Table 50–Contd...

Sl. No.	Name and Address of the Fish Farmers	Date of Stocking	Species Found	Avg. Weight (gms.)	Avg. Length (cms.)	Survival Rate (per cent)	Production in Kg. per Unit (i.e.1500 Sq. ft.)	Remarks
17.	Bir Sing Lepcha, Pemling	April'06	Katli	210	15	43	144	Fish fed with chicken guts+ variety of herbs
18.	Ongden Lepcha, Pemling	April'06	Katli	195	12	47	139.3	
19.	P.T. Lama, Dalapchand	April'06	Katli	170	12	46	152	
20.	Padam Bahadur Chettri, Dalapchand	April'06	Katli	190	14	47	144	
21.	Pratap Sing Rai, Dalapchand	April'06	Katli	240	16	49	160.5	Fish Fed with Chicken guts + variety of herbs
22.	Bhupendra Rai, Dalapchand	April'06	Katli Mahasheer	240	17	56	167	Fed with guts and Beef liver
23.	Shiv Pradhan, Dalapchand	April'06	Katli	160	13	47	139	
24.	Sujita Rai, Dalapchand	April'06	Asala Katli	240	16	56	163	Fed with Domestic waste
25.	Harka Bdr. Rai, Dalapchand	April'06	AsalaKatli	260	17	62	172	–do–
26.	Deo Prakash Rai, Dalapchand	April'06	Asala Katli	200	15	59	163.4	–do–
27.	Kiran Waiba, Lamahatta, Takdah	April'06	Katli	200	13	43	156.53	–do–
28.	Richen Sherpa, Lamahatta, Takdah	April'06	Katli	180	12	54	129.5	Fed with feed
29.	Lekden Sherpa, Lamahatta, Takdah.	April'06	Asala Jatli Mahasheer	190	11	63	173.6	Fed with feed + Domestic waste
30.	Samden Sherpa, Lamahatta, Takdah.	April'06	Asala Katli	170	12	52	162	–do–
31.	Pasang Sherpa, Lamahatta, Takdah	April'06	Katli	190	14	43	139.2	Fed with feed + Domestic Waste
32.	Palwan Sherpa, Lamahatta, Takdah	April'06	Katli	200	16	46	140.9	–do–

Contd...

Table 50–Contd...

Sl. No.	Name and Address of the Fish Farmers	Date of Stocking	Species Found	Avg. Weight (gms.)	Avg. Length (cms.)	Survival Rate (per cent)	Production in Kg. per Unit (i.e.1500 Sq. ft.)	Remarks
33.	Rida Sherpa, Lamahatta, Takdah	April'06	Katli	190	13	37	129.6	Fed with feed + Domestic Waste
34.	Nima. N. Sherpa, Lamahatta, Takdah	April'06	Katli	110	10	42	123.6	Fed with feed
35.	Danima Sherpa, Lamahatta, Takdah	April'06	Asala Katli Mahasheer	150	11	64	167	–do–
36.	S.K. Rai, Okayti T.E., Purana School Dhara	April'06	Asala Katli Mahasheer	280	17	66	182	Fed with Kitchen waste and guts.
37.	Mangan Rai, OTE, Tersing	April'06	Katli	260	15	46	152	–do–
38.	Rudra Mani Pradhan, OTE, Purana School Dhara	April'06	Katli	260	14	41	139.9	–do–
39.	Saran Sharma, OTE, Purana School Dara	April'06	Katli	250	15	43	141	–do–
40.	Sosit Rai, OTE, Purana School Dara	April'06	Katli	240	13	49	153	Report on Production of Coldwater Fishes at Darjeeling as on April
41.	Narbu Gholey, Thurboo T.E., 9th Mile	April'06	Katli	110	12	39	126	Fed with kitchen waste and guts.
42.	Sachi Rai, TTE, Rasdal Goan	April'06	Katli	270	15	44	153	–do–
43.	Moniraj Rai, TTE, Rasdal Goan	April'06	Katli	290	18	46	152	–do–
44.	Tika Ram Rai, Lower Soureni Busty	April'06	Katli Mahasheer	390	17	56	189	–do–
45.	Padam Bdr. Pradhan, Upper Soureni Busty	April'06	Katli	180	14	43	137	Fed with feed
46.	Kesar Rai, Paila Goan, Mirik Busty	April'06	Katli	270	16	42	135	Fed with guts+Kitchen waste waste.

Contd...

Table 50–Contd...

Sl. No.	*Name and Address of the Fish Farmers*	*Date of Stocking*	*Species Found*	*Avg. Weight (gms.)*	*Avg. Length (cms.)*	*Survival Rate (per cent)*	*Production in Kg. per Unit (i.e.1500 Sq. ft.)*	*Remarks*
47.	Shyam Kumar Rai, Paila Gaon, Mirik Busty	April'06	Katli	270	17	49	165	Fed with guts+Kitchen waste waste.
48.	Dayal Chandra Rai, Paila Geon, Mirik Busty	April'06	Katli	290	17	43	149	–do–
49.	Bal Kr. Rai, Paila Goan, Mirik Busty	April'06	Katli	240	16	41	137	–do–
50.	Santa Bdr. Chattri, Siring Gaon, Mirik Busty	April'06	Katli	270	17	44	153.3	–do–

3.20. Prospect of Jhora Fishery

Changes in the biological, chemical, geological or physical component of ecosystems can have impacts on resource population and community. To rejuvenate endangered fish species immediate steps to be taken for mass production of seed and stocking of juveniles in their natural habitat of these rivers and streams. For this purpose there is a need to set up coldwater fish farm in the hill area. It will play a vital role for propagating conservation and culturing endangered species and side by side will also help in producing a healthy commercially important disease free important species that can be raised in their natural habitat. This will also generate new hope for the near extinct important sport fishes. The fisherman of the locality who is dependent on the fish catch will also be benefited. There is also tremendous scope for augmenting the fish production from the Jhora fish pond by manipulating fish combinations of Indigenous species like mahseer (*Tor putitora*), snow trouts (*Scizothorax* sp), Katli (*N. hexagonalipes*), Common carp (*Cyprinus carpio*), Grass carp (*Ctenopharyngodon idella*). In few pockets of Darjeeling areas the farmers have already started using the above combinations, not in scientific manner yet results shows increase in fish production. Hence, if scientific fish culture is adopted for the coldwater fishes, the Jhora fishery scheme will be lucrative venture for the local residents.

Government of West Bengal with the help of D.G.H.C. has set up fish seed farm at Reyang in Kalimpong sub division to cater the demand of fish seeds mainly grass carp and common carp to the fish farmers of the hills and to take up preliminary study of the coldwater fish species as to their captive breeding, food and feeding habit etc. There is an urgent need to start up with the scientific approaches toward the assessment of present fishery status of rivers and Jhoras. Importance is also to be given to the thorough study of biodiversity, water quality, and physico-chemical parameters of different streams to come up with suitable recommendations for the future ventures.

3.21. Extension for Improvement of Jhora Fishery

Practices of Jhora fishery is in vogue in the hill streams of Darjeeling district since time immemorial under traditional system. As a result, the production obtained from such water bodies could not uplift the socio-economic status of the farming communities. During 1986 the Department of Fisheries, Government of West Bengal and Darjeeling Gorkha Hill Council (DGHC) have given prior attention and thrust towards enhancement of fish production through scientific management techniques. But unfortunately the scientific packages of practices have not yet been percolated to the ultimate users to the extent it should have been. Thus, Government have taken initiative to uplift the knowledge, skill and attitude of them, through appropriate extension methods namely organizingation of training programme, group discussion, demonstration and fish farmers' days.

3.22. Conservation of Jhora Fishery

To achieve the sustainable fishery in the hill region it is highly essential to conserve the Jhora resources and the fishery therein. The practices of upland constructive of Jhora may cause severe damage in the area causing soil erosion vis-à-vis landslide. Again mass destruction of fish seed, juveniles in the hill streams through use small meshed net, poising use of explosives etc has negative impact on the development of Jhora fisheries vis-à-vis production. It is necessary to adopt the conservation measures for the growth of sustainable fishery. Training facilities are also inadequate and there is basically no research-training-farming linkage. Therefore, the need for scientific management is vital for maintaining the potentialities of Jhora fishery. Proper knowledge of management will not only lead to the socio-economic upliftment of the beneficiaries but also, lead to sustainable production of the Jhora fishery.

3.23. Future Strategies for Jhora Fishery

1. The department has programme to increase the number of Jhora pond in the hill region of West Bengal by another 4000 numbers within the next two years and thereby more and more poorer people can be involved in Jhora Fisheries activities.
2. More importance would be given to identify indigenous un-conventional species for culture in the Jhora Ponds.
3. For regular supply of seed of culture species all out effort would be given to breed the culture species in the existing hatchery at Reyang.
4. Massive awareness programme would be launched to generate awareness among Hill people regarding culture of hill fishes. There is proposal to impart training to interested farmers at the block level.
5. There are above 4000 nos. of Jhora fisheries unit under the FFDA, those are to be promoted as *"Integrated Farming unit"* coupled with Horticulture along the embankments, to follow energy recycling; and space utilization, thereby minimization of farm economics.

Chapter 4

Commercially Important Culturable Fish Species of North Bengal

Amongst 131 fish species identified from Darjeeling and Jalpaiguri districts, only 21 are presently reported from Darjeeling uplands (Barat *et al.*, 2005). According to Mukherjee *et al.* (2005), about 125 fish species have been recorded from Darjeeling hill area, some of which are found in torrential streams and are remarkably well adopted for clinging to rocks in swift currents. From a zoo-geographical point of view the fish fauna of this area deserves special interest as the area happened to be the ancestral meeting place of the Chinese, Malayan and Indian elements of fishes of the oriental region. Certain special hill stream, Chinese and Malayan, fishes of this region are not found in the Western Himalayas, but it is remarkable that allied forms, sometimes identical are found in the Western Ghats and the connected hill ranges of peninsular India. There is considerable faunastic evidence to show that at some, not very remote, geological period the Eastern Himalayas or the Hills of Assam had a connection with the Western Ghats through the intermediation of Satpura trend of mountains which served as a highway for the migration of torrential fishes from this area westwards and southwards.

The Fishes (Plates 65–74) of Darjeeling Hill streams can be classified under two groups *viz.* (*i*) Non-migratory and (*ii*) Migratory. The non-migratory Fish Species are those which remain in the system permanently and are available throughout the year, such as *Schizothorax sp*, *Nomaeohius sp*, *Glytothorax sp*, *Pseudochensis sp*, and many others. The migratory Fishes are available in the river Teesta only during certain parts of the year *i.e.* from the end of February to October. The important species are

Plates 65–74: Different Fish Species Stocked in Jhora Fish Pond in North Bengal

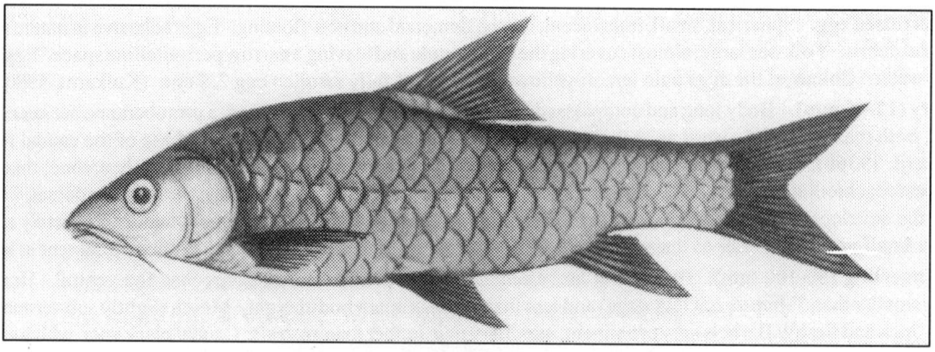

Plate 65: ***Tor tor*** **(Hamilton, 1822)**

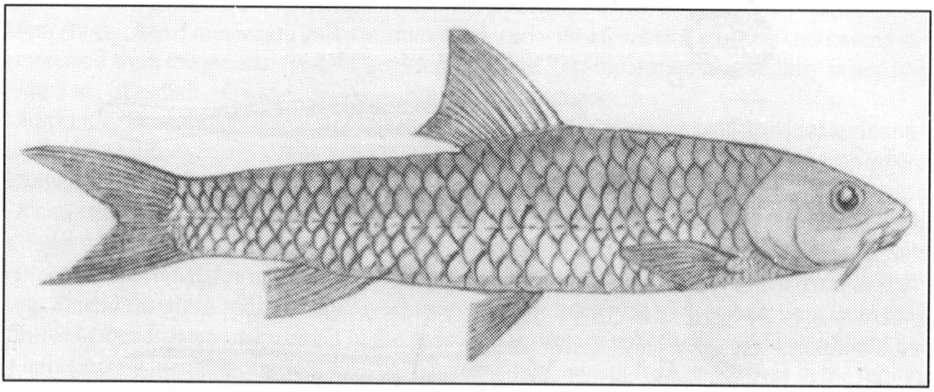

Plate 66: ***Neolissocheilus hexagonolepis*** **(Mc Clelland, 1974)**

Plate 67: ***Cyprinus carpio*** **(Linnaeus, 1758)**

Plate 68: *Ctenopharyngodon idella* (Cuvier and Valenciennes, 1844)

Plate 69: *Catla catla* (Hamilton, 1822)

Plate 70: *Hypophthalmichthys molitrix* (Bleeker, 1959)

Plate 71: ***Labeo rohita*** **(Hamilton, 1822)**

Plate 72: ***Cirrhinus mrigala*** **(Cuvier, 1917)**

Plate 73: ***Labeo gonius*** **(Cuvier, 1817)**

Plate 74: *Labeo dero* (Cuvier, 1817)

Mahseer, *Bagarius bagarius*, *Clupisoma montana* and *Angucilla bengalensis*. They migrate up the river Teesta from the plains of North Bengal and they ultimately enter into Teesta and Rangeet rivers mainly for spawning. Various indigenous species of Fishes of minor importance dominate the different types of water resources of Darjeeling district. The non-migratory fish species are those which remain in the system permanently and are available throughout the year, such as *Schizothorax sp.*, *Nomaeohius sp.*, *Glytothorax sp.*, *Pseudochensis sp.*, and many others. The major fish fauna of economic importance reported from these uplands includes Katli (*Neolissocheilus hexagonolepsis*), Mahseer (*Tor putitora* and *T. tor*), Goonch (*Bagarius bagarius*), Gurdi (*Labeo dero* and *L. dyocheilus*), Indian trout (*Raimas bola*), Asla (*Schizothorax richardsonii*) whereas the fish fauna of less commercial importance comprise Budena (*Garra gotyla*), Termas tera/Utta (*Labeo pangusia*), Faketa (*Barilius bendelisis*), Taktaka (*Barilius barna*), Lohari (*Crossocheilus latius*), Cheepti (*Semiplotus semiplotus*), Bhitti (*Danio aquipinnatus*), Nipati (*Danio dangila*), Titkobri (*Euchiloglanis hodgarti*) and Titakaberi (*Balitoria brucei*) etc. Based on growth performance, maturation, fecundity and breeding some of the indigenous and exotic species are selected for commercial culture in the *Jhora* ponds. These are golden Himalayan mahaseer (*Tor putitora*), (*Tor tor*), Katli, (*Neolissocheilus hexagonalepis*), Snow trout (*Schizothorax richardsonii*), Khursa bata (*Labeo dero*), Gold carp (*Carassius auratus*), and Common carps (*Cyprinus carpio*), Grass carps (*Ctenopharyngodon idella*), Silver carps (*Hypophthalmichthys molitrix*), and Catla (*Catla catla*), Rohu (*Labeo rohita*), Mrigal (*Cirrhinus mrigala*). For better adoption of fish species, fish farmers should know the behavior of the fish biology of important cultivable fish to enhance fish productions from *Jhora*. Thus, biology of some fishes in brief is given below.

4.1. Golden Mahaseer

(*i*) Common Names

Putitora mahseer (Golden Mahaseer), Tor, Mahseer, Mahasol (Bengali); Sor-machha, Saar (Nepalese).

(*ii*) Scientific Name

Tor putitora (Hamilton, 1822)

(*iii*) General Characters

Body strongly built, oblong, somewhat compressed, stream-lined, both the profiles arched gently and gracefully. Abdomen rounded. Head broadly pointed anteriorly and the tail behind the anal fin becomes considerably narrow. Head is always greater than the depth of the body. Mouth small, horizontal with the opening obliquely directed upwards, gap not extended to below the eyes. Puntitora mahaseer has sexually dimorphism. Male and female sex can be separated by the following secondary sexual characters developed during breeding season. Male are smaller and lighter than of the same age. Depth of body in ripe female is about equal to the length of head, whereas in ripe males the head length is smaller than the depth of body. Pectoral fins are rough in males and smooth in female. The length of pectoral fin in both the sex is found to be largely variable, and thus, this character is not taken as a dependable one. Anal fin is smaller than pectoral and opaque in female; whereas in male the anal fin is equal or sub equal to pectoral and more transparent. A conspicuous growing body streak appears in males during breeding season but it is absent in females. The streaks originate in front of the lips of the maxillary bone and extend posteriorly along and below the eye-ball up to opercular bone on both sides of the head. Males are brighter than general appearance. Anal fin of female is brighter than that of male. However, the colour characters are not much reliable since these are variable due to different conditions of the eco-system. The putitor mahseer is believed to be the biggest dimensions among the mahseers and has been seen to reach 2710 mm (9ft) in length, weighing over 200 pounds (90.9 kg).

(*iv*) Distribution

"Yellow fined", "Grey-hound", or common Himalayan "Mahaseer" occurs all along Himalayas. It is also reported from China. It was first recorded in Keshi hills. It was established in Uttar Pradesh, western Himalayas, West Bengal, Assam, Punjab, and eastern Himalayas. Outside India it is found in Pakistan and Bangladesh.

(*v*) Habitat

T. putitora is inhabitant of both lotic and lentic environments of coldwater streams and lakes. The greatest abundance of its population appears to be in the mountain streams of rocky base. Its general availability is in the major rivers and their tributaries at a height between 750 m and 1300m, having slow current and deep pools, which are sometimes chocked with filamentous algae and submerged vegetation. Fishes of the size 3-5 kg are normally available in these deep pools. It remains both in the shallow and the deeper waters; the smaller fishes up to few inches in length are normally found in any clear shallow and gravelly steams in the Duars and tarai; and larger the mahseer run. Fry hide themselves under big boulders near the banks; juveniles prefer the midstream's where the current is faster, hiding beneath stones and often running in shoals; and the adults normally remain in the deeper pools of the main streams.

(*vi*) General Biology

Puntitor mahseer is the premier sporting fish of India, but little is known of its general biology. It is a very strong and hardy fish, and it can stand long-distant transport. Its strongly built streamlined body and powerful locomotory organs provide quicker movement against the torrential current of the hilly streams. It is a local migrant fish. It performs normally seasonal migration for breeding, feeding or wintering. Water temperature and floods determine the limits of such migrations. Putitor has nest-making habit, which they do in the form of depression on the river bottom before their spawning.

(*vii*) Food and Feeding Habits

The Mahaseer is noted to be intermittent feeder and information on the food and feeding habits is scanty. The species has more choice for animal-food in adult stage but juveniles have preference for foods of plant origin and detritus. It feeds on green filaments, insect larvae, algae, water plant, and slimy matter from rocks and hard substratum. The detro-plankton feeding habit of juveniles is verily confirmed by the suctorial mouth facilitating bottom feeding.

(*viii*) Reproduction

The species attains its first maturity in the length of 330-360 mm and in 1.5 to 2 years. Putitora breed in the north-western Himalayas thrice a year, once in winter (January-February), next in summer (May-June) when the river are swollen with snow-melt water, and lastly in the monsoon (July-October) when the hill streams are flooded due to south-west monsoon. Putitor mahseer performs the spawning migration from the main rivers to tributaries and breeds in the flood phase. They lay eggs on amidst stones, pebbles, gravels, etc., in the shallow areas of the streams in the water depth 30-40 cm. The artificial breeding has been successfully done by 'wet stripping methods' and 'dry stripping methods' where fertilization of eggs was 50-80 per cent and 80-100 per cent respectively.

4.2. Turiya Mahaseer

(*i*) Common Names

Tor mahseer (Turiya Mahaseer), Tor, Mahasol (Bengali); Saar (Nepalese).

(*ii*) Scientific Name

Tor tor (Hamilton, 1822)

(*iii*) General Characters

Body stoutly built and elongated, dorsal profile more sharply arched than ventral, abdomen rounded. Head is shorter than body. Snout is not tuberculated. Mouth terminal small; gape does not extend to below the eyes; but horizontal with the opening obliquely directed upwards. Lips are thick, fleshy with continuous labial fold. Body colour usually silvery or grayish green along the dorsal half, becoming silvery shot with gold on the sides and beneath; lower fins reddish yellow or grayish green dorsal, with sides of body in the middle of a pinkish colour which is replaced by greenish gold above and light olive green below; fins usually orange. *Tor tor* is heterosexual. Sexual differentiation is made by secondary sexual characters during breeding seasons. Males are smaller and lighter than female of identical age. Body is narrower in males in comparison to that in female. Head length-body width relation in male is higher than in female during breeding season. Bulkiness of the abdomen for enlarged roes gives rise to an arched ventral profile in female. In male the abdomen is almost straight and the milt oozes on slight pressure on the abdomen. Pectoral fins in male are rough in touch and comparatively longer reaching up to 6th lateral line scale. In female it is smooth and it reaches to the 5th scale.

(*iv*) Distribution

Tor tor is widely distributed along with foothills of western Himalayas, east Punjab, Uttar-Pradesh, Bihar, Darjeeling district of West Bengal, Assam, eastern Himalayas, Ambala district of Haryana and Madhya Pradesh of India, Pakistan, Bangladesh and China.

(*v*) Habitat

Tor tor normally inhabits cold freshwater of both lotic and lentic environments– the streams, lakes and reservoirs of the lower reaches of the Himalayas which receive snow-melt water directly from water-sheds; and in some upland waters of the Decan plateau. They prefer torrential streams with oxygenated water and rocky pools with gravel. Their populations are generally found in the sub–mountainous algae at elevations below 1300 m of the substratum of stones, pebbles, gravels and sand under tropical/warm climate having temperature range 4-30°C. Their natural occurrence appears to concern with the water temperature prevailing in the streams than the altitude.

(*vi*) General Biology

Tor tor possesses a stream-lined body with very large and powerful fins. The size of the fins together is greater in area than the total superficial area of the rest of their body, which gives a strong power in locomotion through the torrential current of hilly streams. For their extensive feeding habits on underwater rooted vegetation and algae in natural habitat, they are considered as a very promising agent for biological control of submerged weeds. Mahseer builds nest by scooping gravel in rocky terrain. They perform migration in a limited range for the purpose of breeding, feeding or wintering. They determine the limit of such migration by water temperature and floods. The morphological variations recorded in the morphometric and meristic characters of *Tor tor* from different eco-environments/locations appear to be phenotypes. The fertilized eggs and the newly-emerged hatchlings remain stuck to the under surfaces of stones, pebbles, gravels, etc., to escape from being swept away by the strong current of the streams.

(*vii*) Food and Feeding Habits

Tor tor is a bottom feeder and herbivorous in diet subsisting mainly on the macro-vegetation and also consumes insects, mollusks and fish matters to some extent. The food organisms of juveniles and adults are same but the items constituted in varying degrees. Insects formed the main food of the juveniles, while macro-vegetation was the maximum in the adult. The insectivorous feeding habit of juvenile are gradually changed to herbivorous dietary habits of adults. Adult are rather carnivorous to some extent, feeding on mollusks and insects.

(*viii*) Reproduction

Tor tor attains maturity by the end of the first year of life in lentic environment and end of the 2nd year of life in lotic environment. Mature Males are relatively smaller in size. Tor tor is polygamous and the mating seems to be promiscuous. The mating pair exhibits some nesting habit in the breeding activity and mating intensity in short. *Tor tor* does not breed in confined water of ponds, though it spawns in lakes and its connecting streams, where the fluvetile conditions prevail. However, the successful artificial breeding has been achieved by stripping method.

4.3. Chocolate Mahseer

(*i*) Common Names

Chocolate mahseer, Snub-nosed mahseer (Copper Mahaseer), Bhorkol, Katli, Buluk (Bengali); Katli (Nepalese)

(*ii*) Scientific Name

Neolissocheilus hexagonolepis (Mc Clelland, 1974)

(*iii*) General Characters

Body elongated and compressed. Depth of body almost equals to length of head. Dorsal and ventral profiles equally and slightly convex. Abdomen rounded. Interorbital space is rather convex. Snout is obtusely rounded. Several rows of horny tubercles on the sides of snout and sub orbital region are found. Body colour olive green superiorly, silvery white below, bluish tinge on dorso-lateral region, with a golden yellow lateral band above lateral line; edges of the scales marked with light bluish netural tint; tip of snout green; two oval patches of a light yellow colour before and behind the eyes; irish yellowish brown; gill cover light alizarin pink; dorsal fin citron green, while pelvic, anal and greater part of pectoral and caudal slate-gray. A patch of buff colour is on the pectoral. *Neolissocheilus hexagonolepis* is heterosexual. Sexual differentiation is made by the help of the secondary sexual characters which are developed during breeding seasons. The roughness of the pectoral fin in males is not always a definite character. Similarly, colouration of the body in both the sex is variable in different eco-systems and it is not therefore considered as a firm character for sex identification.

(*iv*) Distribution

Neolissocheilus hexagonolepis is widely distributed along with foothills of eastern Himalayas, Darjeeling district of West Bengal, Assam, Meghalaya, arunachal Pradesh of India, Nepal, Bangladesh, Burma, Thailand, Malaya, Sumatra and Siam.

(*v*) Habitat

Neolissocheilus hexagonolepis (Katli) normally inhabits cold freshwater of both lotic and lentic environments–the streams, of the eastern Himalays. Katli is belived to be a freshwater fish at home in hill streams and small and large rivers, but also suitable for cultivation in confined waters. This species is found to be well established in the terraced ponds of Darjeeling district hill areas, where it breeds in the confinement under the influence of water flow artificially created into the ponds *i.e. Jhora* fishery.

(*vi*) General Biology

The species name hexagonolepis is ascribed for the hexagonal shape of the exposed portion of the scale. Katli is a beautifully coloured fish and colouration may vary considerably to the nature of the water inhabited by them. They possess transversely convoluted intestine and longer fins for powerful locomotion against the strong current of the hill streams, but do not make the traffic rush mahseer. They rather bore down into the depth of the water and shake the line like bull dog.

(*vii*) Food and Feeding Habits

Neolissocheilus hexagonolepis is an omnivorous feeder and it has bottom feeding habit near the marginal shallow. Being a voracious eater, its stomach and gut are usually gorged with food. Gastropod shells, filamentous and planktonic algae, vegetable debris, and sand or mud are the main food items encountered in the stomach. The major items of food in fry stage are animalcules and water fleas. In the early fingerling stage, the insect larvae, aquatic beetles and flies are mainly consumed; whereas the aquatic macrophytes and marginal grass from the main food of advanced fingerlings and adults.

(*viii*) Reproduction

The sexually maturity in Katli is attained fairly early in life and at the end of the first year. Much disparity is seen in the sexual attainment of maturity. Male matures when they are still relatively much smaller in size as compared to the mature females. Katli is polygamous. The mating pair exhibits some nesting habit in the breeding activity and mating intensity in short. *Neolissocheilus hexagonolepis* do not breed in confined water of ponds, though it spawns in lakes and its connecting streams. However, the successful artificial breeding has been achieved by stripping method.

4.4. Snow Trout

(*i*) Common Names

Snow trout, Asala (in Nepali)

(*ii*) Scientific Name

Nakkatua -*Schizothorax richardsonii,* Chuche -*S. richardsonii*

(*iii*) General Characters

Schizothorax richardsonii is characterized by the presence of a suctorial mouth. The length of the head and the high of the body are almost equal and are contained 4-6 times in the total length. The mouth is slightly oblique, horse-shoe shaped and the upper jaw is longer. The maxilla reaches to nearly below the front edge of the eye. Lower labial fold is interrupted in the middle. The lower jaw is lined internally with a horny covering. The burbles are four in number. The dorsal fin originates slightly nearer the base of the caudal fin than to the end of the snout. The caudal fin is deeply forked. Body-colour is bluish along the black, becoming silvery with numerous golden spots, most distinct in the both side of the body.

(*iv*) Distribution

Generally the snow trout species are of central Asian origin. It is distributed in Himalays, from the head waters of the Teesta and Rangeet rivers of Sikkim. It is also reported from Nepal, Bhutan and Southern Tibet.

(*v*) Habitat

Normally inhabits cold freshwater environments–the streams, of the eastern Himalays.

(*vi*) Food and Feeding Habits

It feeds on small insect larvae as well on the scum covering the stones.

4.5. Common Carp

(*i*) Common Names

Common carp, European carp, German carp, Carp, Karp, American rui (Bengali)

(*ii*) Scientific Name

Cyprinus carpio (Linnaeus,1758)

(*iii*) General Characters

Body is elongated, stout, moderately compressed, both the profiles arched; abdomen rounded and broad. Head moderate, snout blunt, rounded. Mouth terminal and protrusible; gape narrow; lips thin and continuous, upper lip with 4 long barbells. Branchial openings are rather wide, reaching beyond hind border of opercle. Gill membranes broadly united with isthmus. Dorsal fin long, origin opposite to ventral, and extend up to opposite mid-anal. Caudal fin well forked, its lobe being pointed at the tip. Colour much variable; back usually blue-green to dark, brownish or olive brown; sides blue-green to golden yellow; under surface pale to whitish; iris pale; dorsal and caudal dull brownish, other fins pale to whitish, often toned with red.

(*iv*) Distribution

Common carp is the only fish which has perhaps a world-wide distribution since the current century. There are anomálous views about the origin of the wild carp. The carp is a native of the temperate parts of Asia and especially of China. A consignment of common carp was imported from Bangkok and stocked in Cuttack waters in 1959 for culture and propagation.

(*v*) Habitat

The original habitat of common carp is a river fish in the region of the earth, which is characterized by a hot summer and a cold winter, scarcity of rain; light seasonal, with the altitude generally over 1000 m above sea level. Being widely distributed in the tropical and temperate region and through extensive selective hybridization between imported and the local verities of the carp, the strains evolved have been found to be acclimatized to a variety of habitats and extreme of environments. Domesticated for centuries, it is found to be equally suitable in small and large ponds, reservoirs and lakes and in rice fields. Common carp forms the most common fishery of the world. Many countries have extensive and intensive farming of carp. Common carp makes the most acceptable warm-water as well as coldwater fish for commercial fish culture. Carp culture often faced with heavy destruction of stock at various stages of lives.

(*vi*) General Biology

Common carp is a hardly fish. It has extremely good adaptability in variable external conditions, and thus, it has been domesticated nicely. Fair range of tolerance for water temperature and salinity is of good quality for carp to culture in the pond. The carp has the inherent capacity to withstand even serious injuries and regenerate its wounded parts quickly. The carp has no set pattern of migration. In open water it performs local migration. In enclosed bodies of water carp fry and fingerlings may exhibit schooling in shallows, and in larger specimens this habit may not be obvious. Common carp has a low to medium sustenance ratio, the ratio of protein, carbohydrate and fats requirement for catabolic activities. The food conversion ratio of the carp is markedly high.

(*vii*) Food and Feeding Habits

Common carp is a polyphagous or omnivorous, and it has wide food spectrum. It can adjust the dieting habit very nicely according to local availability of natural food. Although the carp is not usually a plankton feeder in the adult, but they are sometimes found feeding voraciously on zooplankton if it is abundant in the environment. It has in general preference for bottom flora and fauna. The feeding intensity of the carp is varying with the variations in the water temperature. Feeding is almost stopped at 3°C, but becomes 19 per cent of the normal at 4°C, some food is consumed in winter at 4-8°C water temperature, intensive feeding being when temperature rise to 14°C, and the most intensive feeding appears at 24-28°C. The food of common carp of Asian varieties as Chironomidae, Ephemeridae, Crustacea, Mollusca, Trichoptera and farinaceous feeding stuff. Carp do also accept rather easily the artificial food of various types, such as dries Mysis, tubuficid worms, freshwater gastropod, vivipara corn, viscera of birds and mammals, dries fish, soyabean cake etc.

(*viii*) Reproduction

Common carp is heterosexual. Sex identification in immature fish is very difficult. On the approach of spawning season the secondary sexual characters are developed, when male is easily recognized by the milt which excused freely from deep pit-like genital pore on application of gentle pressure on the abdomen. Females show swollen/ bulging soft abdomen and prominent strongly protruded reddish fleshy vent with its distal margin gently cleft at the middle. The ripe males sometimes have large bulging abdomen. The right testicle is usually smaller than the left, so is the case right hard roe. Sexually maturity in carp depends upon the biological factors (age, size, sex) and the prevailing climate conditions (especially temperature) in which the fish is grown. The maturation of carp is faster in tropics. Common carp, a seasonal spawner in temperate water, breeds normally once in a year in wild conditions during spring, when the temperature is increasing, but the cultivated stock may breed twice or more under exceptionally good conditions. The "wild spawning" of common carp in ponds, rivers, reservoirs and lakes is mostly controlled by the ambient water temperature.

4.6. Grass Carp

(i) Common Names

Grass carp; White amur; Gheso –Rui (Bengali); Grass carp (Nepalese)

(ii) Scientific Name

Ctenopharyngodon idella (Cuvier and Valenciennes, 1844)

(iii) General Characters

Body elongated, oblong, both the profiles gently arched, abdomen narrowly rounded. Head broad with short round snout. Upper jaw is slightly longer and protractile than the lower jaw distinct only at angle of mouth. Mouth is slightly subterminal and oblique. Lips are thick and simple. A labial fold runs along the upper lip. Barbel is absent. Body colour is dark grey, with greenish shade in live condition, on the dorsal, fading to light grey on the sides and silvery on the belly. The base of each scale is dark brown. All fins dark. Fins are somewhat rounded. Dorsal fin originates a little advance of the ventral base.

(iv) Distribution

Grass carp is a native to China and USSR, occurring in the lower and middle reaches of the river Amur, Ussuri and Sungari and in the Lake Khanka, on the Pacific coasts and in the rivers Liao, Hai, Yellow and Min along southern regions of China. The present global distribution of the species is for its successful transplantation to about 60 countries, where the fish has been established well through farm culture. In India, it was first brought in 1959 from Hongkong and Japan. The gift consignment of grass carp fry were sent from India to Philippines in 1964 and to Nepal in 1966.

(v) Habitat

Originally the grass carp was riverine stock but through intensive introduction and accidental release this becomes a regular inhabitant of reservoirs, lakes, tanks, ponds etc. Though it is a freshwater species but it survives in brackish water up to 9 per cent salinity. It is a highly adaptable species and can tolerate wide range of physico-chemical conditions. They can survive at temperature down to 0°C to 40°C or above. The natural environment of adult and young grass carps must be with sufficient aquatic vegetation to meet their voracious feeding habit. They prefer to be in the littoral zone where weed population is relatively more in deeper water bodies, but in shallower ponds they frequently move all the regions.

(vi) General Biology

This carp is well known for its insatiable appetite for weeds, thus becomes a most suitable agent for biological control of excess weed from the pond. The wide environmental tolerance, long lifespan and high fecundity are the important bio-physical characteristics permitting them to maintain their populations in nature.

(vii) Food and Feeding Habits

Grass carp at larval stage feeds on plankton with preference to nanoplankton, rotifers and small sized cladocerans. But they quickly switch over their feeding habits on macrophytes entirely. In pond conditions the size at which grass carp change their diet to aquatic plant has not still been established. Young ones prefer softer plants.

(*viii*) Reproduction

Grass carp is heterosexual. In mature male nuptial tubercle is present on head and on dorsum of caudal peduncle. In mature female extrovert, swollen pinkish coloured vent is seen. The average age and size at first maturity are seen to be considerably varied in different countries. In tropical countries the sexual maturity appears at earlier ages and in smaller sizes. Grass carp is polygamous and mating is promiscuous. Fertilization is external. The gonadosomatic index is relatively low when compared with Indian major carps. In nature grass carp normally breeds once in a year, but through hypophysation multiple breeding within a year can be done.

4.7. Silver Carp

(*i*) Common Names

Silver carp, Rupali Rui (Bengali); Silver carp (Nepalese)

(*ii*) Scientific Name

Hypophthalmichthys molitrix

(*iii*) General Characters

Body is oblong, fairly laterally compressed. Abdomen is strongly compressed, with a sharp keel from brest to vent. Head is moderately large, snout bluntly rounded. Mouth sub-superior; the lower jaw with a tubercle rather upturned, and the upper jaw strongly notched; lips thin; cleft not extending to anterior margin of eyes. Eyes are rather small, anterior, subinferior, and visible from below ventral surface. Gill membrane not linked with isthmus. Suprabranchial organ is present. Gill-rakers longer than gill-filaments and joined in the form of spongy sieve-membrane. Single rayed dorsal fin is small in size, originating slightly behind ventrals, inserted midway between tip of snout and base of caudal. Pectoral fins are large reaching the ventral origin or a little beyond of it. Scales are cycloid, thin and minute. Lateral line decurved, but passes through the middle at the caudal peduncle region. The back is black grey and the rest of the body silvery. Fins are dark.

(*iv*) Distribution

The river system of Yangtze, West River, Kwangsi, Kwangtung in south and central China, and the Amur Basin in USSR are the native water for silver carp. India brought 360 fingerlings of silver carp from Japan in 1959, reared in the pond at Cuttack (Orissa), and hypophysed seed was distributed as gift consignment to Philippines in 1964, Nepal in 1966 and Burma in 1968. In India, silver carp is well established through farming, but accidental escape in the Govind Sagar reservoir made its natural distribution in the system. This species was also introduced to Pakistan and Bangladesh; in Sudan; and in many other European waters.

(*v*) Habitat

Silver carp is originally a riverine stock, pelagophilic in nature and prefers swift current, but subsequently adopted in lentic environment through domestication. It can tolerate slight brackish water as well. Cold weather, considerably low water temperature and highly loaded phytoplantonic coldwaters do not cause infavourable to them.

(*vi*) General Biology

Silver carp is highly sensitive, much more than the grass carp. Long duration transport of silver carp along with the other carps under oxygen packing causes early mortality of the former. During transport silver carp are very uneasy, get bruised and die very easily. Silver carp has specialized structure of the gill rakers adopted to micro-plankton feeding. Silver carp, especially pond-reared, is susceptible to various types of infections, causing diseases and also leading to death in certain cases.

(*vii*) Food and Feeding Habits

Silver carp is planktophagus. Correlating the feeding habits from zoo to phytoplankton the silver carp shows certain anatomical and morphological changes in the digestive tract of different size groups.7-9 mm long and 1-3 days old fry commence feeding on zooplankton, such as rotifers and nauplius of copepods. Fry and adult feed on *flagellata, dinoflagellata, myxophyceae, bacillariophyceae, protozoa,* and *rotifer* supplemented by decade macro-vegetation and detritus.

(*viii*) Reproduction

Silver carp is bisexual. The characteristic process and ridges appearing on the pectoral fin rays in the males. These ridges are found even in the immature fish, the secondary sexual characters developed during the breeding season are marked as follows for male and female. In male, a row of fine body ctenoid teeth appear on several foremost fin-rays of the pectoral fins, especially on the unbranched rays. These are sharp when touched. Theses ctenoid teeth will not disappear once grown out. Abdomen is rather small, with silt-like genital aperture. White milt flows out when the part of the testes near the vent is gently pressed. The reproductive cycle of silver carp is monsoon oriented, and it follows a circannual rhythm, similar to that of grass carp, catla, rohu and mrigal. The onset of functional maturity of finely reared stock of silver carp is however much earlier, even in summer months of March-April in West Bengal. Silver carp is polygamous and mating seems to be promiscuous. Fertilization is external.

4.8. Catla

(*i*) Common Names

Catla; Katla (Bengali); Katla (Nepalese)

(*ii*) Scientific Name

Catla catla

(*iii*) General Characters

Body is moderately elongated and deep. Head is broad and large. Abdomen rounded. Mouth superior, lower jaw turned upwards, gape very wide, upper lip absent, lower lip moderately thick and not fringed. No barbells. Scales are cycloid and moderately large. Large and broad dorsal fin commencing a little advanced of ventral, 2/3rd as high as the body. Dorsal with 14-16 branched articulated rays. Anal fin is shorter than dorsal one. Body is grayish colour on the back, silvery on the sides and beneath.

(*iv*) Distribution

Catla is a Gangetic carp and its natural distribution extends up to the Godavari river in the Andhra Pradesh of India. In India, catla was more common in northern part but for successful transplantation since the beginning of the 20th century it has now spread over the whole of peninsular region of the country as well as too many other continents. These species have been well established in all those countries forming there a regular fishery of commercial scale.

(*v*) Habitat

Freshwater rivers, reservoirs, lakes, jheels, beels and ponds are the common habitats of catla. Indian major carps (IMC) including catla are found to occur in certain sections of the river Ganga, Yamuna and Sarda in Uttar Pradesh during some part of the year within a limited temperature, 57°F appears to be the minimum temperature tolerable by IMC.Adults are normally surface dweller. Fry and early fingerlings move in shoals preferably along the marginal water.

(*vi*) General Biology

Among carps catla is having comparatively less bones. It has soft flesh while young, but becomes coarse in texture beyond 80 cm. It is the fastest growing fish among the cultivable Indian carps, and domesticated by nature. It generally remains in deep water. Its fecundity is fairly high. It responds well in inducement of breeding in suitable, confinement.

(*vii*) Food and Feeding Habits

Catla is predominantly a planktophagic surface feeder. For plankton feeding habit a great inter-specific food competition is seen in catla with almost all the inhabiting species at fry stage as they all mainly prey upon the plankton. The degree of competition declines on age as other species change their feeding habit in different niches. The minnows and the clupeids are the main competitors of catla. Catla is a surface feeder having preference for crustaceans and algae. Next to these, plant matters form the important food items. Utilising this propensity the thick growth of *Hydrilla, Oscillatoria* and Spirogyra have been successfully recommended for Catla culture.

(*viii*) Reproduction

Catla is heterosexual. Catla attains sexual maturity in the second year of life. Maturity even appeared in 18 months in well managed ponds of fish-farmers in West Bengal with balanced rational feeding and fine rearing practices. The female indulge to meet the males first, and flash and jump out to get rid of males. Afterwards courtrship is made and pairing is done between the spawning female and male. On of the males held the females, bend his body round her abdomen, and knock, rub and nudge her. At the climax of mating the pairs are coiled at the tail region of the body placing their genital apertures close to each other, in such a manner the spawning activity of a pair may continue for 30-60 minutes or even more. The mating intensity of a pair prolongs in cool weather. Catla does not spawn naturally in confined water, but breeding occurs in rivers, reservoirs and in waters providing fluviatile conditions. No single factor is responsible for natural spawning of catla. Catla breeds in confined waters by fish pituitary hormone injection, and successful spawning was first achieved in May, 1959 at Joysagar, Assam, India.

4.9. Rohu

(*i*) Common Names

Rohu, Rui,(Bengali); Rohu (Nepalese)

(*ii*) Scientific Name

Labeo rohita

(*iii*) General Characters

Body is moderately elongate, dorsal profile more convex than that of the abdomen, somewhat concave over the orbit. Snout is obtuse, without labial folds, projecting beyond the jaws. Mouth inferior, gape narrow, lips thick and fringed, and with distinct inner fold to each lip. Teeth are pharyngeal. The maxillary pair of barbles is thin and concealed in lateral groove; the rostral pair, when present, small, and often missed. Scales cyloid and moderate in size. Body colour bluish or brownish along the back, become silvery on the sides and beneath, usually with a red mark on each scales during breeding season(Khan and Jhingran,1975); fins grayish, sometimes black, eyes are reddish.

(*iv*) Distribution

The distribution of rohu is in the freshwaters of Sind, and from the Punjab through India and Assam to Burma, and is not found in Madras or on the west coast.

(*v*) Habitat

The natural habitat of rohu is the freshwater rivers, reservoirs, lakes, 'jheels' 'beels', and ponds, below an altitude of approximately 540m. Adult fish prefers to dwell in deep waters, while the larvae remaining surface or sub surface waters. Young and mature ones wander in the midwaters of pond or in the shallower region of the deeper waters for feeding and breeding purposes.

(*vi*) General Biology

Being a column feeder herbivore with quick growing nature in terms of flesh return, rohu is considered to be a very useful compatible member of 6-species composite fish culture of Indian and Chinese carps to produce maximum yield. Rohu produce maximum eggs per unit body weight among the three main species of Indian major carps.

(*vii*) Food and Feeding Habits

Rohu is predominantly a surface feeder and planktophagus at fry stage. For selective zoo and phytoplankton feeding habit of fry, rohu forms a great interspecific food composition with almost all other inhabiting culturable carp species at earlier stages of life. But this competition is considerably reduced from fingerling stage onwards when rohu change the feeding habit and habitat, become a column-bottom feeder, and start consuming food mainly the filamentous algae, rotten vegetables, and sand and mud.

(*viii*) Reproduction

Rohu is heterosexual. Male and female sex recognized only when the former oozes milt on slight pressure on abdomen and the later shows buldge soft abdomen with extrovert swollen reddish vent, during breeding season. Rohu normally attains maturity in the second year of life in Indian environment. Rohu is polygamous and it seems to be promiscuous. The mating activity of rohu appears to be the same as that of catla. Fertilization is external. Rohu has very high egg-producing capacity among the three Indian major carps. The pond reared well-fed stock produce more eggs than the wild ones.

4.10. Mrigal

(*i*) Common Names

Mrigal; Mrigal (*Bengali*); Mrigal (*Nepalese*)

(*ii*) Scientific Name

Cirrhinus mrigala

(*iii*) General Characters

Body elongated, abdomen rounded, body-depth and head length nearly equal, snout depressed and obtusely rounded. Mouth terminal, gape transverse and wide, lips without fringe, upper lip not continues with lower, lower lip sharp and with a knob at the mandibular symphasis. Single rayed median dorsal slightly nearer to snout than caudal base. Pectoral is shorter than head. Dorsal fin height lesser than body depth, anal not extending to caudal, caudal homocercle deeply forked two rostral barbles short. Scales cycloid, large and 40-45 rows found on the lateral line series. Body colour usually dark grey on the back and silvery on the sides and beneath; fins greyish, tips of pelvic, anal and the lower lobe of caudal tinged orange, especially during breeding season.

(*iv*) Distribution

The natural distribution of mrigal is in the freshwaters of northern India, Bangladesh, Burma, and Pakistan. It was introduced into south Indian waters where it has been established well. Alongwith other Indian major carps mrigal has also been successfully transplanted to Nepal, Malaysia, Srilanka, Japan, Mauritus, Philippines, South Rodesia, U.S.S.R., Africa, Vientien, etc. from Calcutta since 1957 onwards and is reported to have been established well in some of those countries.

(*v*) Habitat

Freshwater rivers, reservoirs, 'jheels', 'beels' and tanks are the natural habitat of mrigal. It is an excellent species for pond culture. Mrigal is a bottom dweller. Adult fish generally frequent in the deeper water although they move in the shallower region for feeding and breeding purposes. Spawn and fry inhabit mainly the surface and sub-surface water.

(*vi*) General Biology

Mrigal serves as an important fishery in Bengal and some other north-eastern states of India, and shares a good production in the commercial catches of the river Ganga, Yamuna and Brahmaputra and their tributaries. *C.mrigala* is a very fast growing big sized cyprinid. For herbivorous feeding habit and other cultural qualities mrigal has been extensively utilized as a nice compatible member of six species composite fish culture in practice in India. For culture some weed growth is better as the adult mainly subsists on blue-green and filamentous green alga, pieces of higher aquatic plants, decayed vegetable matter, mud detritus, etc. Mrigal hatchlings consume artificial feed better than by that of rohu and catla.

(*vii*) Food and Feeding Habits

Mrigal is predominantly a bottom feeder in adult stage, but surface and mid-water feeder in fry and fingerling stages. Although it is an illiophage in feeding habit and stenophage in food variety, but its mouth has a typical characteristics of a planktophagus, *i.e.*, the mouth is large and immobile. Mrigal fry commences feeding on the third day after hatching, when the alimentary canal becomes functional. Fry mainly subsists on zooplankton. Adult feeds mainly on blue-green and filamentous green algae, diatoms, pieces of semi-rotten higher aquatic plants, decayed vegetable matter and detritus. In general, the feeding was comparatively poor in the adult than in the young ones.

(*viii*) Reproduction

Mrigal is heterosexual. Male and female sex can be differentiated only during breeding season, when oozes milt on slight pressure on the abdomen near the vent and the female shows bulged and soft abdomen with extrovert swollen reddish vent. Mrigal normally attains sexual maturity in the second year of life in Indian environment. Male matures earlier than the female. The pond-reared well-fed stock of both the sexes is, however, found to be matured by 15 months in West Bengal waters. Fertilization is external. Mrigal grows well both in the lentic and lotic environments and performs better growth in the old tanks with adequate vegetable debris.

4.11. Kursa

(*i*) Common Names

Gonius; Kursa (*Bengali*); Kursa (*Nepalese*)

(*ii*) Scientific Name

Gonius (*Labeo gonius*);

(*iii*) General Characters

Body elongated and compressed with rounded abdomen, dorsal profile more convex than ventral. Snout obtusely rounded and covered with a number of prominent tubercles. Mouth is inferior, gape narrow, lips thick and fringed without lateral lobes and with a distinct inner fold along with the entire circumference. A cartilaginous horny covering is supporting inner side of both the jaws without teeth. Teeth on pharynx are in rows of 5.4.2./2.4.5. Barbels 4, rostral pair shorter than the maxillary.

Scales are cycloid, very small. Single median dorsal is with articulated rays. Pelvic abdominal is with rays. Body coloured light greenish along the back becoming lighter along the side and the abdomen. 11-13 lateral longitudinal light stripes running from behind opercle to caudal, majority of scales with reddish mark on centers, the marks being prominent on mid-lateral sides.

(*iv*) Distribution

Throughout India, Bangladesh, Burma, Nepal and Pakistan have natural distribution for *L.gonius*. It's distributed as 'Indus in Sind', throughout the north-west provinces, Bengal, and Orissa to Ganjam, as low as the Assam and Burma.

(*v*) Habitat

L. gonius is an inhabitant of the major freshwater rivers, reservoirs, lakes, 'jheels', 'beels', and tanks. It is a carp of plains, common in Sutlej and Beas rivers (Khan, 1944), and also found in coldwaters as high as Kumayan Himalayas. In Mahanadi, though found in both rapid and slow moving waters, they prefer to live in the shallower part of the river where there is plentiful growth of submerged weeds.

(*vi*) General Biology

L. gonius is the slowest growing fish among Indian major carps. Gonius has less demand for quick perishability of flesh, insipid taste, numerous small intermuscular bones, and non-availability of pure seed in adequate quantity in all regions. Gonius is not so hardy in nature and can not tolerate salinity. The information on the parasitism of this species is lacking.

(*vii*) Food and Feeding Habits

L. gonius is a herbivore and predominantly a bottom feeder. In feeding habits, gonius makes a keen competition with mrigal, calbasu, fimbriatus, and bata. The vegetable matter and the sand and mud are the chief food in adult stage. Larvae and postlarvae lives upon mainly on the cladocerons and rotifers. The feeding activity is found decreasing with the advancement of gonads towards full maturity. Males appeared to be more active in feeding than females in entire year. The highest feeding activity has been noticed in adult fish from the lakes of Udaipur region in March and least in July.

(*viii*) Reproduction

L. gonius is heterosexual. The sexual differentiation for male and female is only possible during maturing in March-April onwards, on the appearance of secondary sexual character. Like other Indian major carps, the roughness on the pectoral fins and on the scales, oozing of milt on slight pressure on the abdomen near vent, relatively longer pectorals, etc. of milters and smoothness of pectoral and bulging abdomen with extrovert swollen vent of spawners are the common identifying characters. The usual age of maturity in *L. gonius* is three years like other carps but maturity in two years is also not without exception. The mating activity in gonius, as observed from the natural breeding by hypophysation in confinement, appears to be same as seen in the other major carps. Fertilization is external. The process of syngamy and the cytological events of the developing embryo are same as in other carps.

4.12. Kursa Bata

(*i*) Common Names

Kursa bata; Kursa bata (*Bengali*); Gardi/Thant (*Nepalese*)

(*ii*) Scientific Name

Labeo dero

(*iii*) General Characters

Body slender, dorsal profile more convex than ventral, head oval and blunt, abdomen rounded. Snout is with more or less deep groove across it and often with some pores. Mouth inferior, gape narrow; lips continuous, a horny covering to inside of lower jaw. Tip of mouth with an indistinct lateral lobe, often having pores. Barbells 2 pairs, short maxillary pair often missed. Dorsal fin originates above last quarter of pectoral fins. Caudal deeply forked, upper lobe sometimes slightly the longer. Body colour is dull silvery grey being darkest along the back. Scales cycloid often marked red. Dorsal and caudal fins grey tinged with red dots, other fins salmon coloured.

(*iv*) Distribution

The natural distribution of *L. dero* is all along the Himalayas, in Nepal, Pakistan, Sind hills, Burma and China. In India, its distribution extends from Arunachal Pradesh through Assam, North Bengal, Uttar Pradesh, Himachal Pradesh, Eastern Punjab, and Jammu and Kashmir, particularly in the foot region of the Himayan range. The river Sutlej, Beas, Ravi, Baner, Tawi and Jhelum abound with this species.

(*v*) Habitat

L. dero is a common and more abundantly available coldwater fish inhabiting the upland rivers, streams and brooks. Like the mahseer, it is True Mountain carp preferring the fast flowing channels and rocky pools.

(*vi*) General Biology

Labeo dero is of the Bangana kind (Hamilton, 1822). It is a minor carp and forms a commercially important food fish in the upland waters. The narrow gill-opening and the reduced gills suit their dwelling in the oxygen-rich torrential streams and rivers. The mouth bounded by thick lips and lined internally by a cartilaginous base is highly suitable for their feeding habit of attached herbs from the hard substratum. By nature the kursa bata is hardier performing swift locomotary function in the torrential current, but can not stand even little salinity for exclusively freshwater physiological constituents of the body.

(*vii*) Food and Feeding Habits

L.dero is a herbivour. The soft aquatic vegetation and periphyta are the normal food. The later items are removed by them by well modified lips. Their ill-adopted mouth is unsuitable for feeding in deep and muddy waters. The short and mascular esophagus; long and coiled intestine; well developed intestinal bulb; and thin, almost hair-like gill-rakers partially adapted for sieving the microscopic plant food, are the characteristics of the kursa bata being a typical harbivour. Fishes are first settled on rocks, apply mouth closely to the substrata and move round the rocks till the slimy growth polished off from stones. During feeding the lower lip, which is free from the jaw, is perhaps turned over and the papillae of the dorsal surface of the lower lip come in contact with the substratum.

(*viii*) Reproduction

L. dero is heterosexual. Like other carps the secondary sexual characters of kursa bata are developed and become prominent during pre-spawning and spawning season, when males are recognized by the roughness on the pectoral fin surface and with the oozing milt on slight pressure on the abdomen, and the females by the distended and bulging abdomen with extrovert, swollen, pale pinkish genital organ. *L. dero* attains sexual maturity in the second year of life. Ripe females are found in the nature during May-July in India and Nepal. Male mature earlier than female and ripe males are seen in April also. The spent stage arrives rather quickly for short breeding season. *L. dero* is a polygamous fish and the mating seems to be promiscuous. Both the sex migrate upland stream for breeding. A large numbers of males and females move together and congregate at the breeding ground. Sex matching is possibly held during spawning migration, and mating is taken place at the shallower regions of the breeding ground. *L. dero* is a slow growing carp.

List of common fish species of the rivers of North Bengal is depicted in Table 51. The major fish fauna of economic importance reported from different parts of North Bengal are shown in Plate 75-138.

Table 51: List of Common Fish Species of the Rivers of North Bengal

Local Name	Scientific Name	Comment
Raja baam	*Anguila bengalensis*	Teesta (Sevoke Plain)
Katli	*Neolissocheilus hexagonalepis*	All most all tributaries of Teesta
Gooch /Limbuni	*Bagarius bagarius*	Ghish/Teesta Milan Pally (Jalpaiguri)
Titakabri/Titay	*Balitewra brucei*(Gray)	Teesta (Sevoke Plain)
Puti	*Barbus conchonius*	Reyang/Teesta (Sevoke Plain)
	B. ticto	Reyang/Teesta (Sevoke Plain)
	B. stigma	Reyang/Teesta (Sevoke Plain)
Boroli/Chalay	*Barilius bendelisis* (Hamilton)	Teesta (Sevoke Plain) and near Teesta baradge
	Barilius barna	Reeyang
	Barilius vagra(Hamilton)	
Sole	*Channa striata*	Balasun/Sevoke Plain
Taki	*Channa punctata*	Odlabari/Ghish
Hile	*Channa stewartii*	Thirsat Khola
	*Chana amphibious**	* (Critically endangered)

Contd...

Table 51–Contd...

Local Name	Scientific Name	Comment
Jal Kapur	*Clupisoma Montan**	Upper Rangeet and near Teesta bazaar * (Critically endangered)
Lohari/Utta	*Crossocheilus latius*	Teesta (Sevoke Plain)
Chepti	*Cyprinion semiplotum**	Rishi, * (Critically endangered)
Bhitti/Punti/Naptee	*Danio devario*	Reyang/Relli/Ghish
	Danio dangila	Thirsat khola, Teesat (Sevoke Plain)
	Danio aquipinnatus	Thirsat Khola/Tar Khola/Rangeet
Til Kaprey	*Euchiloglanis hodgarti*	Teesta/Rangeet/Reyang/Rishi
Kaprey	*Pseudocheneius sulcatus*	Ghish/Rishi/Reyang/Rongit
Buduna/Nak-katwa	*Garra gotyla*	Teesta/Rishi
	Garra annandalaiei	Reyang/Relli/Tarkhola
	Garra lampta (Hamilton)	
Bele	*Glossogobius giuris*	Balasun, Teesat (Sevoke Plain)
Kaaray	*Glytothoray peetionptenus*	Teesta/Rangeet
Gardi	*Labeo dero*	Teesta/Sankosh/Ghish/Jaldhaka
Lahari	*Labeo pantusia.*	Teesta (Sevoke Plain)
Jhinge macha	*Macrobranchium sp.*	Sevoke Plain
Teramas Tera	*Mastocembalus armatus*	Balasune plain, Teesta (Sevoke Plain)
Asala-(Snow trout)	*Oreinus molesworthii*	Mainly Rangeet/Rishi
Naak katwa	*Schizothorax richardsoni* (Gray)	Teesta
Chucche	*Schizothorax progastus* (Mell)	
Chanda	*Parambassis nama*	Sevoke
Kholsa	*Polyacanthus fasciata (Colisa)*	Sevoke/Mirik lake
Garela	*Schistura inglisi* (Hora)	Sevoke
	Schistura beavani	Ghish
	Lepidocephalichthys sp	Reyang
	Acanthocobitis botia	Relli
	Aboricthys elongates	Palla
	N. rupelii	Ghish
Pahari pabda	*Silurus cochinchinensis*	Thirsat Khola/Reyang
Paheli sar or Mahasol/Golden Himalayan Mahaseer	*Tor putitora*	Teesta/Rongit/Relli/Reeyang
Red finned Mahaseer	*Tor Tor*	Teesta/Rongit/Relli/Reeyang

Plate 75: *Barilius* sp.

Plate 76: *Barilius vagra* (Boroli)

Plate 77: *Barilius barna* (Boroli)

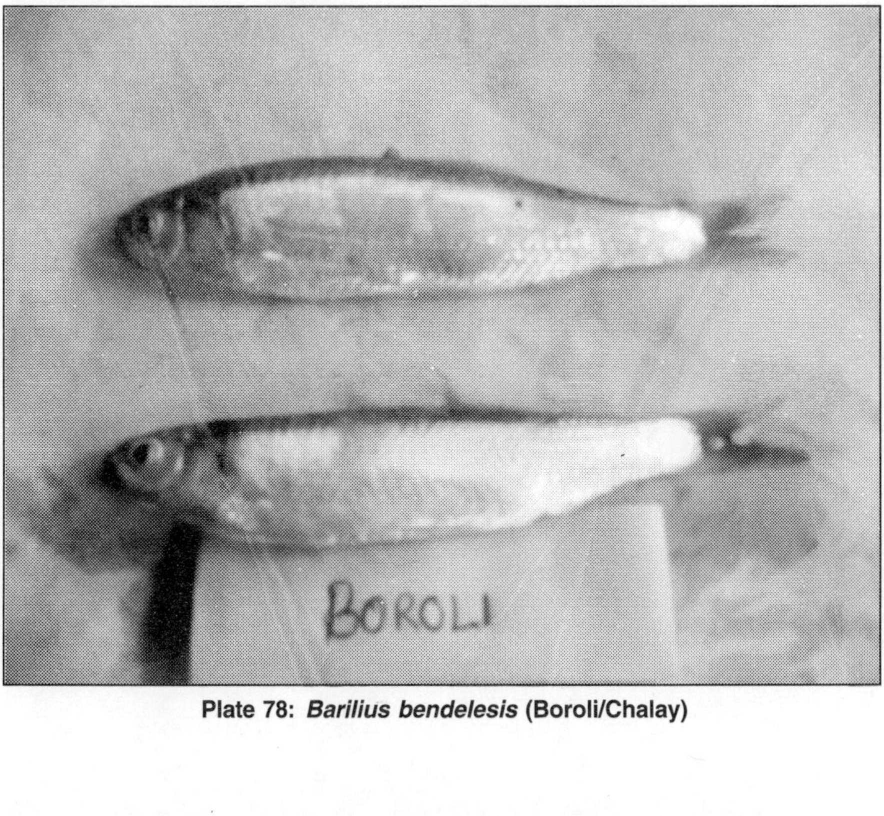

Plate 78: *Barilius bendelesis* (Boroli/Chalay)

Plate 79: *Schizothorax richardsoni* (Asala)

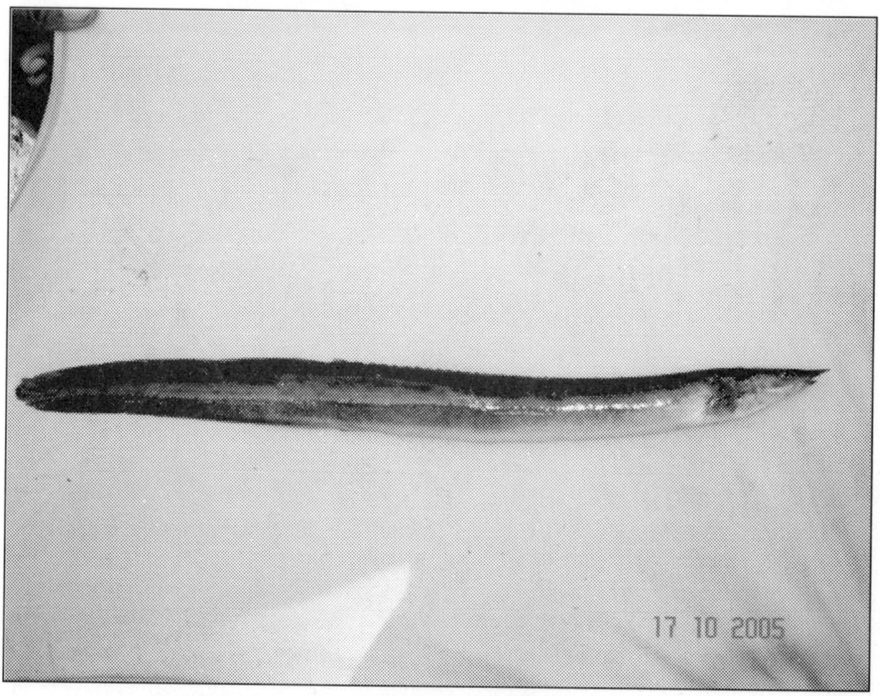

Plate 80: Chuche Bam

Plate 77: ***Barilius barna*** **(Boroli)**

Plate 78: ***Barilius bendelesis*** **(Boroli/Chalay)**

Plate 79: *Schizothorax richardsoni* (Asala)

Plate 80: Chuche Bam

Plates 81–82: *Angula bengalensis* (Raja Bam)

Plate 81

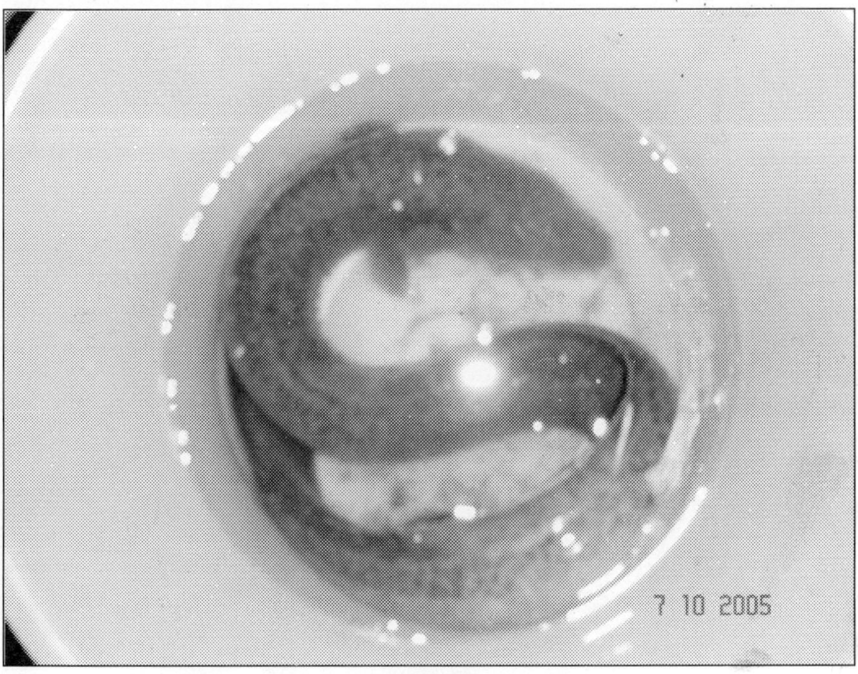

Plate 82

Plates 83–84: *Glossogobius giuris* (Bele Fish)

Plate 83

Plate 84

Plate 85: ***Cyprinion semiplotum*** **(Chepti)**

Plate 86: ***Labeo dero*** **(Gardi)**

Plate 87: ***Neolissochilus hexagonolepis*** (Katli)

Plate 88: ***Tor putitora***

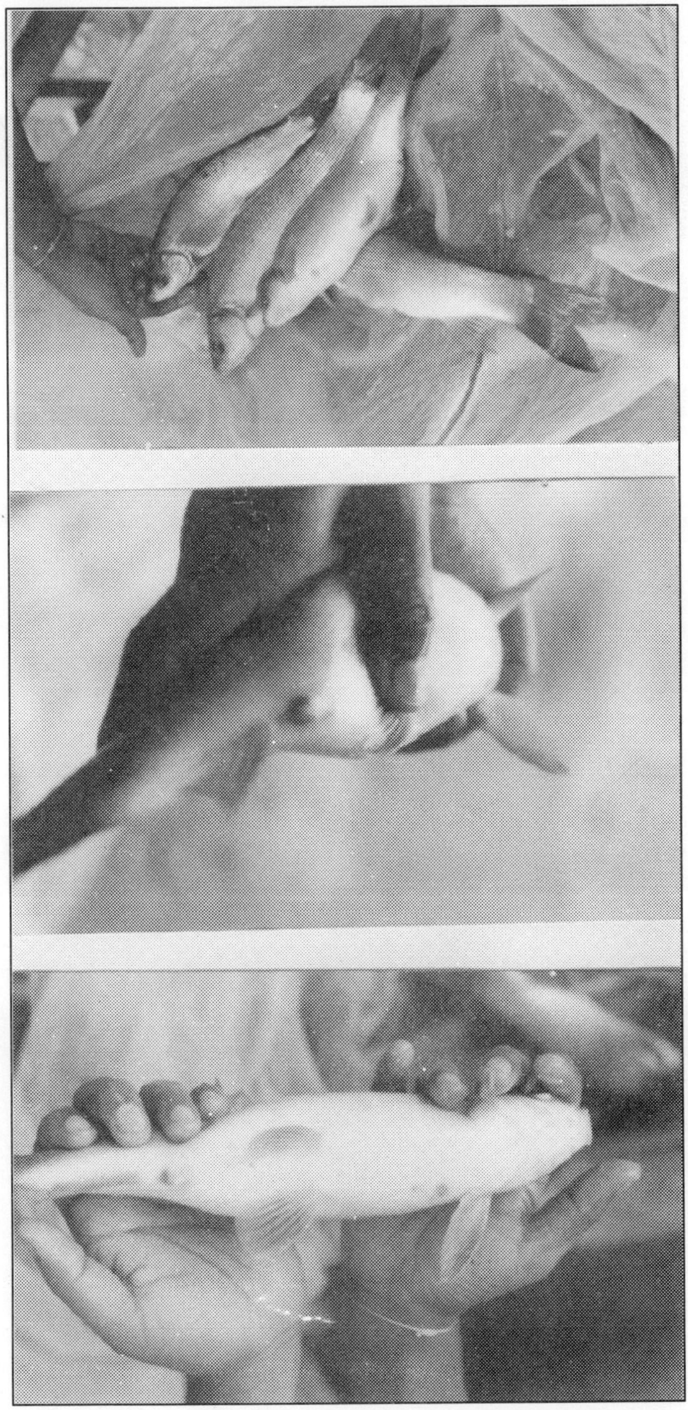

Plate 89: *Labeo gonius* (Kursa)

Plate 90: ***Barbus ticto*** **(Punti)**

Plate 91: ***Barbus*** **sp. (Alta punti)**

Plate 92: *Barbus sarana* (Punti)

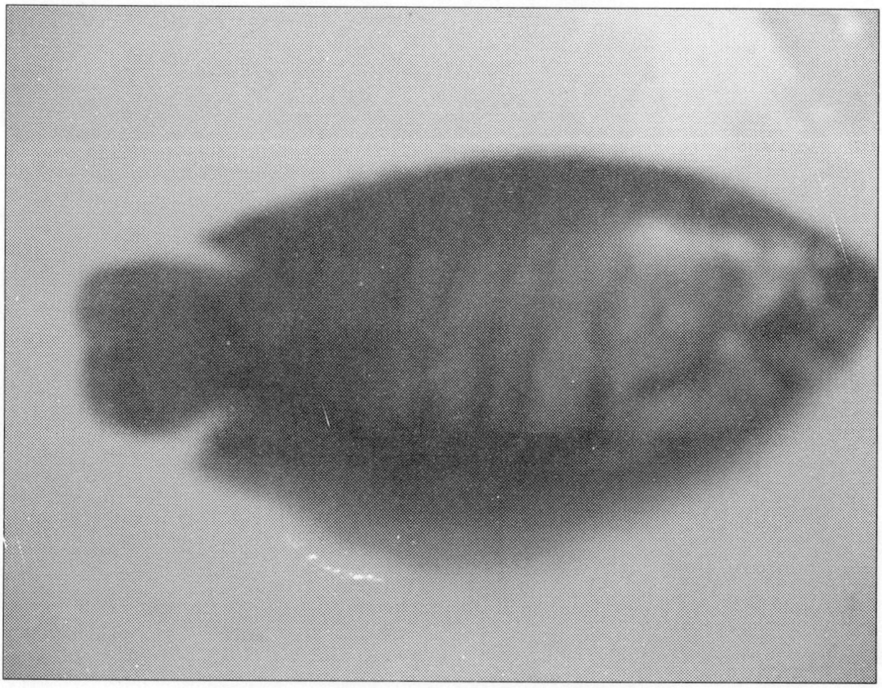

Plate 93: *Polyacanthus fassiata* (*Colisa*)

Plate 94: *Garra gotyla* (Buduna)

Plate 95: *Crossocheilus latius* (Lohori)

Plate 96: ***Garra lamta*** **(Buduna)**

Plate 97: ***Danio aquipnnatus*** **(Bhitti)**

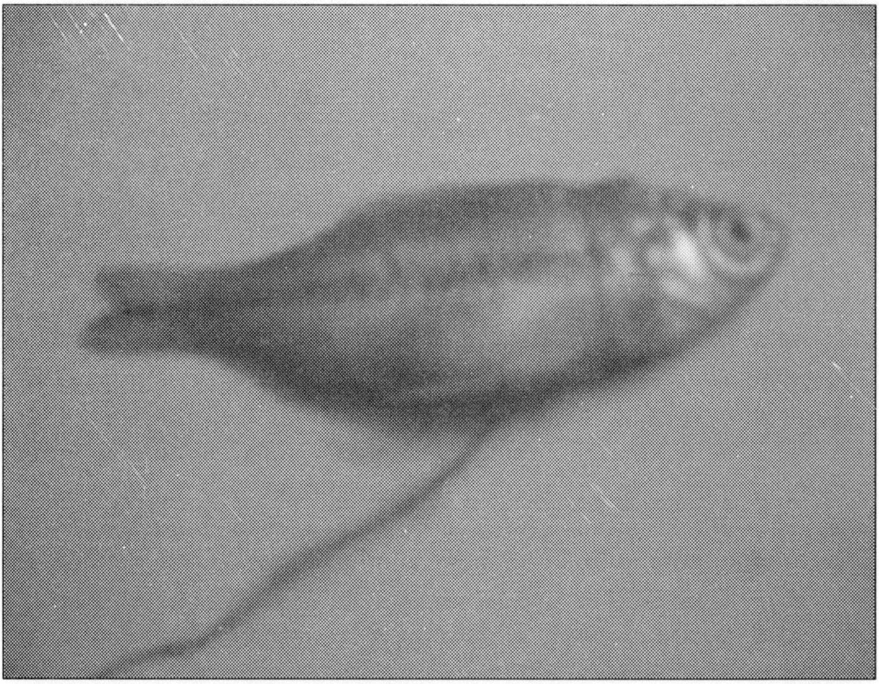

Plate 98: ***Danio devario*** **(Bhitti)**

Plate 99: ***Botia dario***

Plate 100: *Botia lohachata*

Plates 101-104: Varieties of *Botia sp.* (Loach)

Plate 101

Plate 102

Plate 103

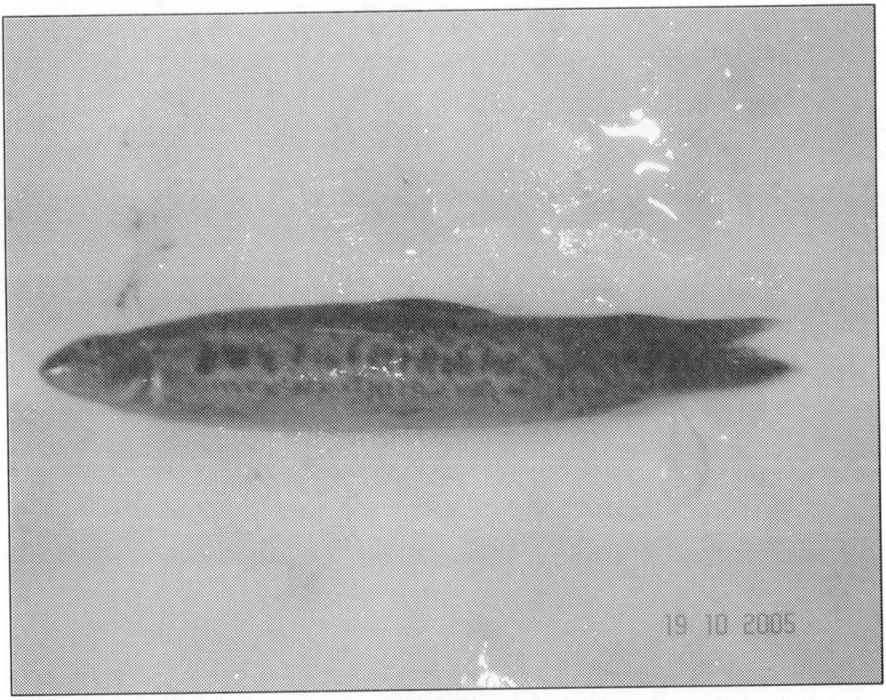

Plate 104

Plates 105–109: Varieties of Hill Loaches

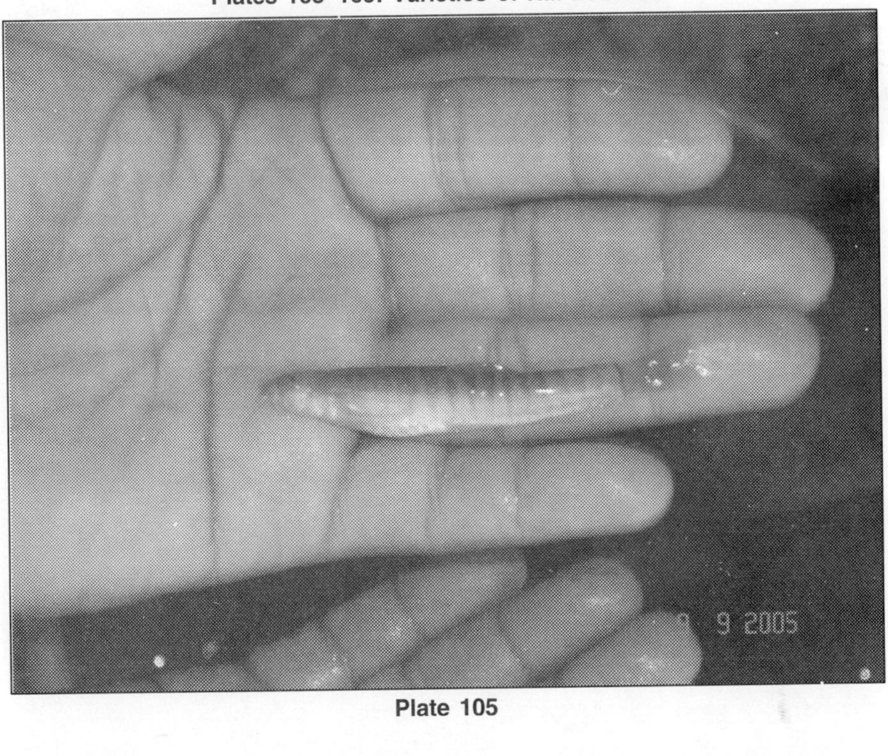

Plate 105

Plate 106

Plate 107

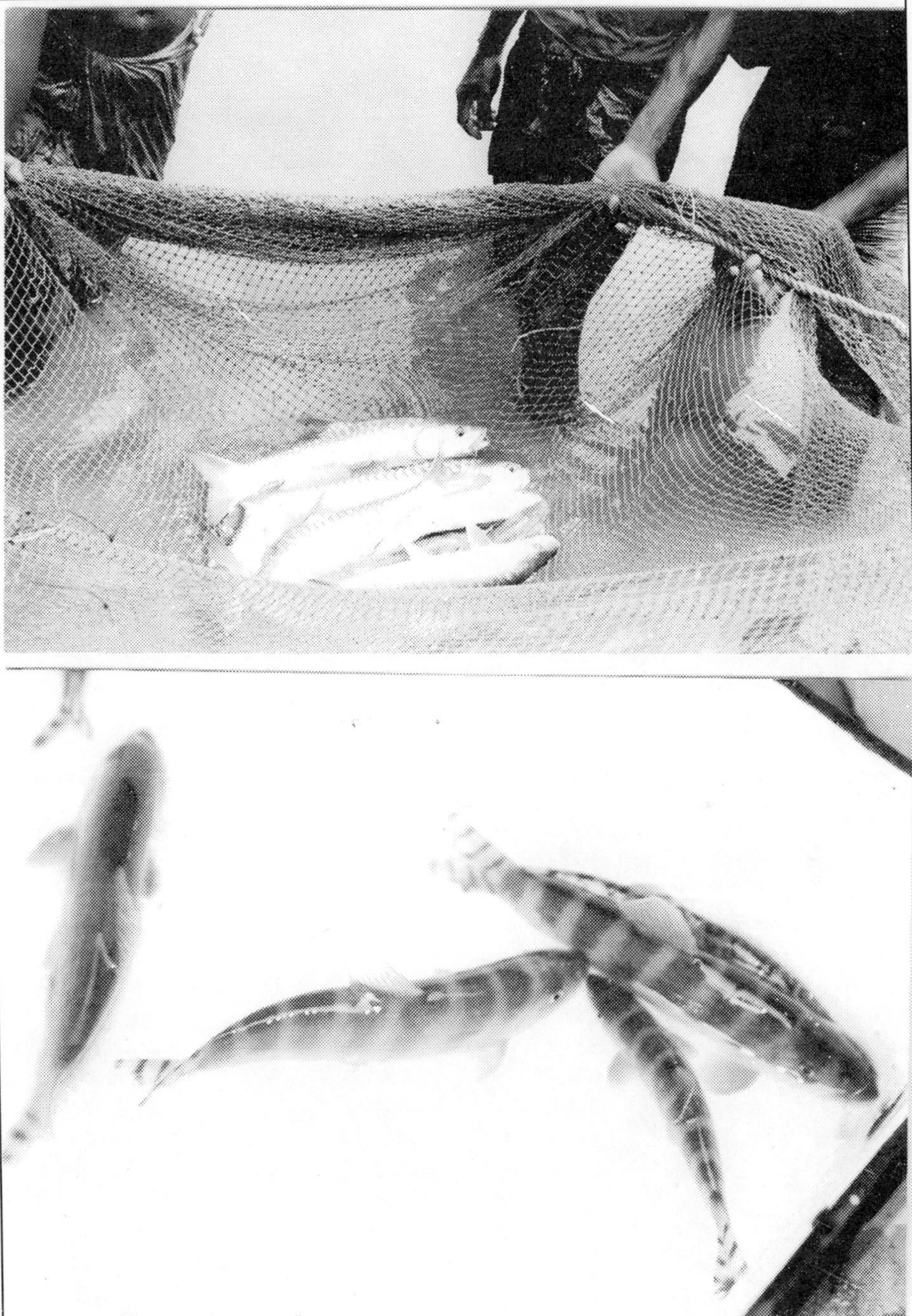

Plate 108

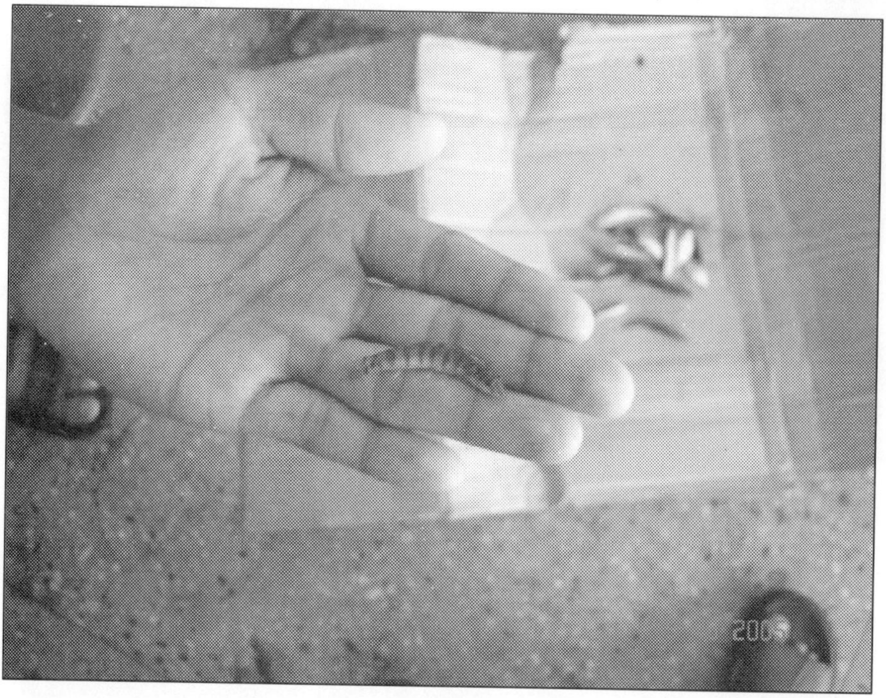

Plate 109

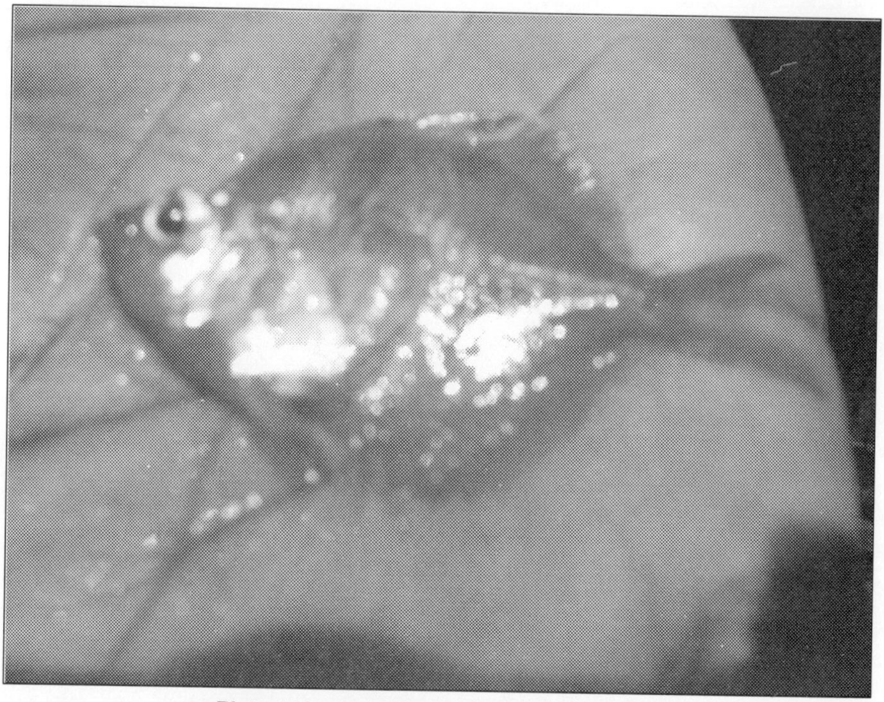

Plate 110: ***Parambassis ranga*** **(Chanda)**

Plates 111–112: *Parambassis nama* (Chanda)

Plate 111

Plate 112

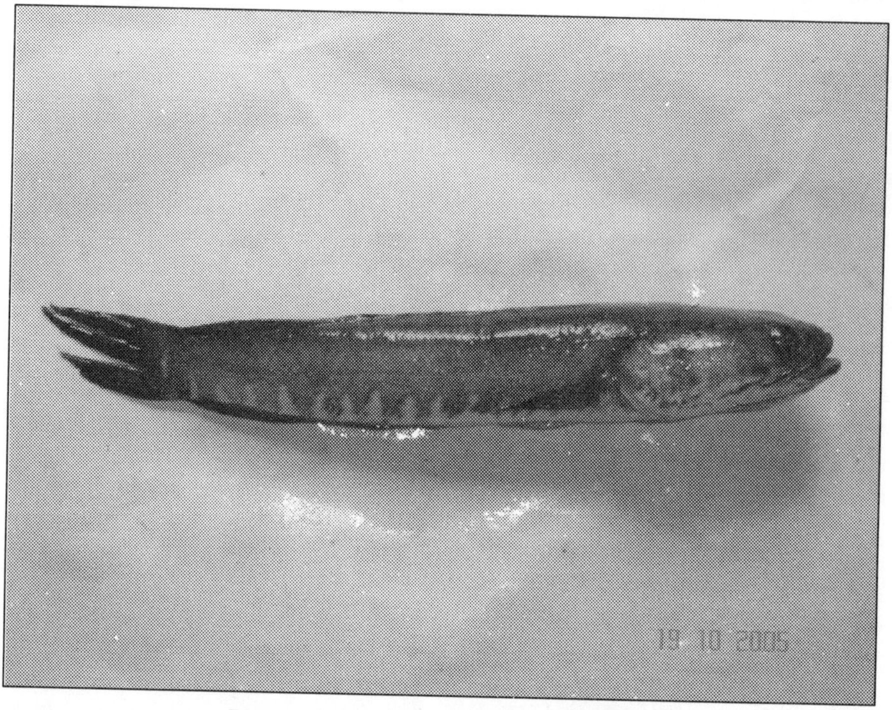

Plate 113: *Channa punctata* (Taki/Hile)

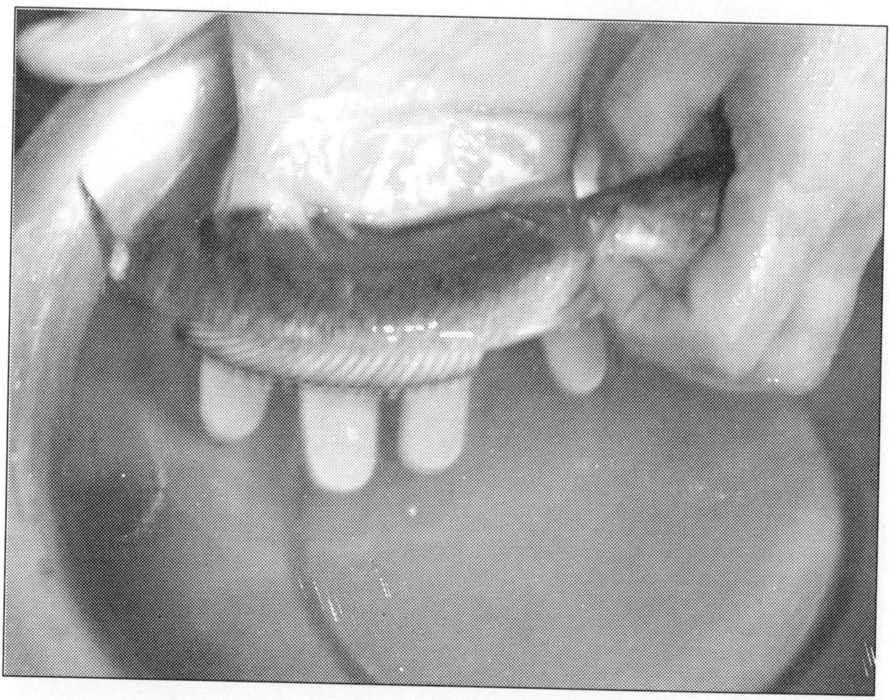

Plate 114: *Channa stewarti* (Taki/Hile)

Plates 115–116*: Balitewra brucei* (Tilkaprey)

Plate 115

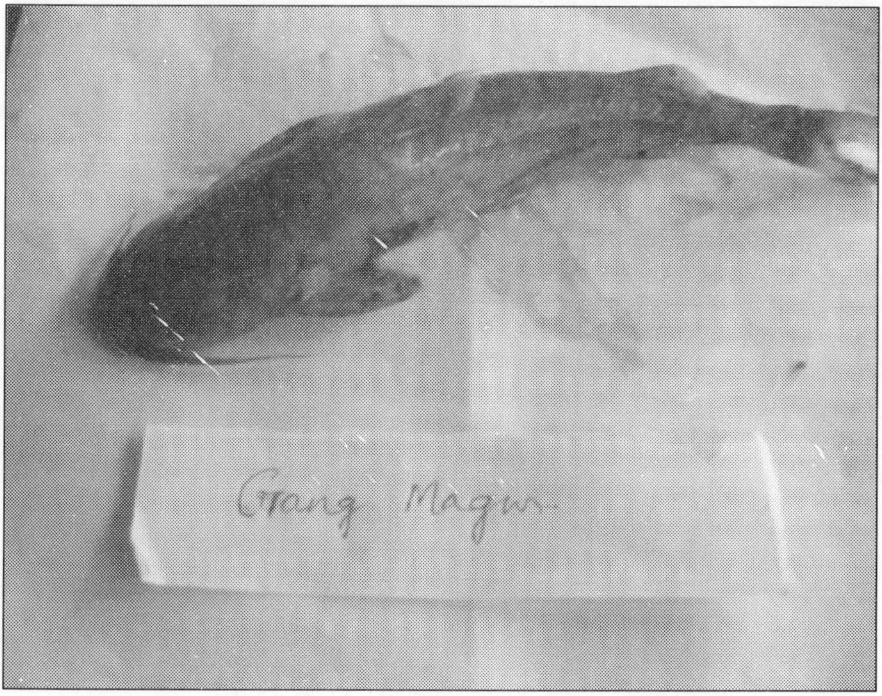

Plate 116

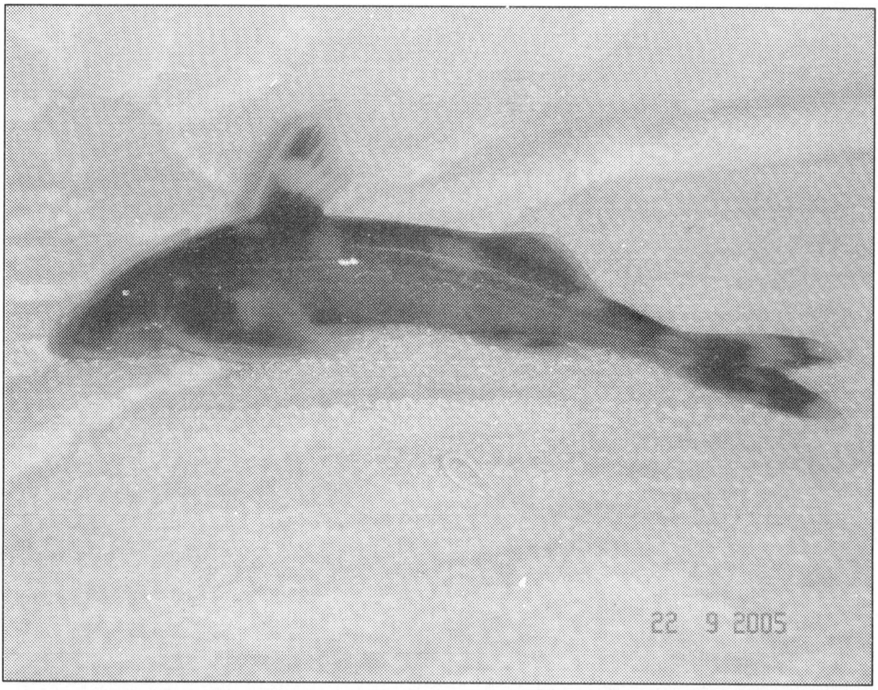

Plate 117: *Pseudocheneius sulcatus* (Kaprey)

Plate 118: *Mystus* sp. (Gochi Tangra)

Plates 119–120: Ventral Side of *Pseudocheneius sulcatus* with Sucker and Without Sucker

Plate 119

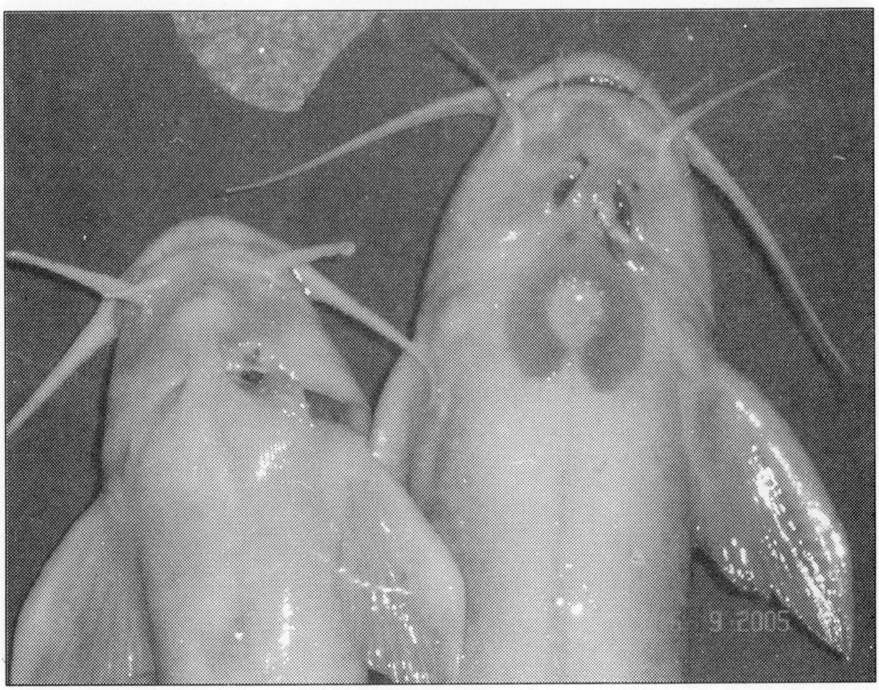

Plate 120

Plate 121: *Mystus cavasius* (Tangra)

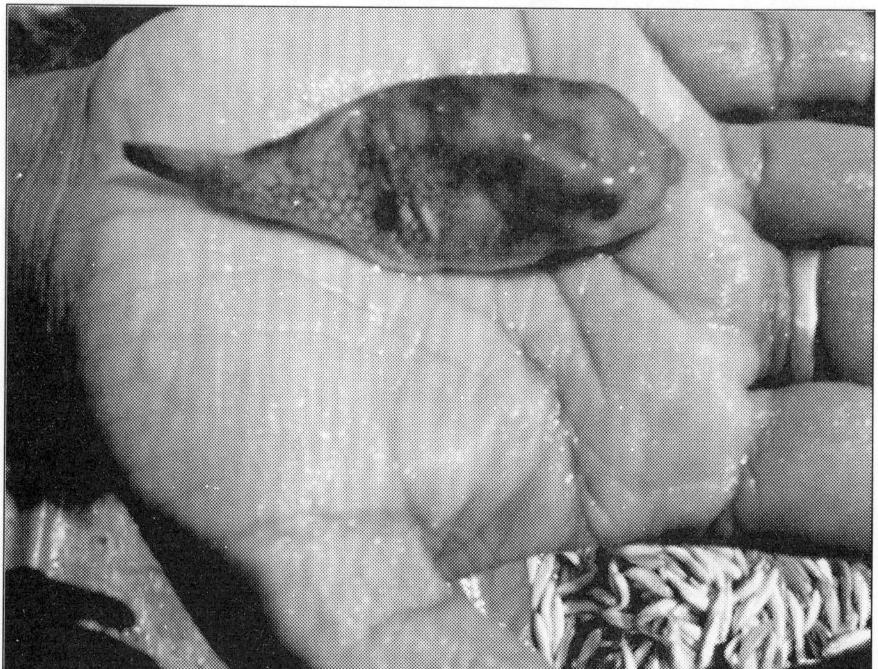

Plate 122: Freshwater puffer

Plate 123: *Ghutum*

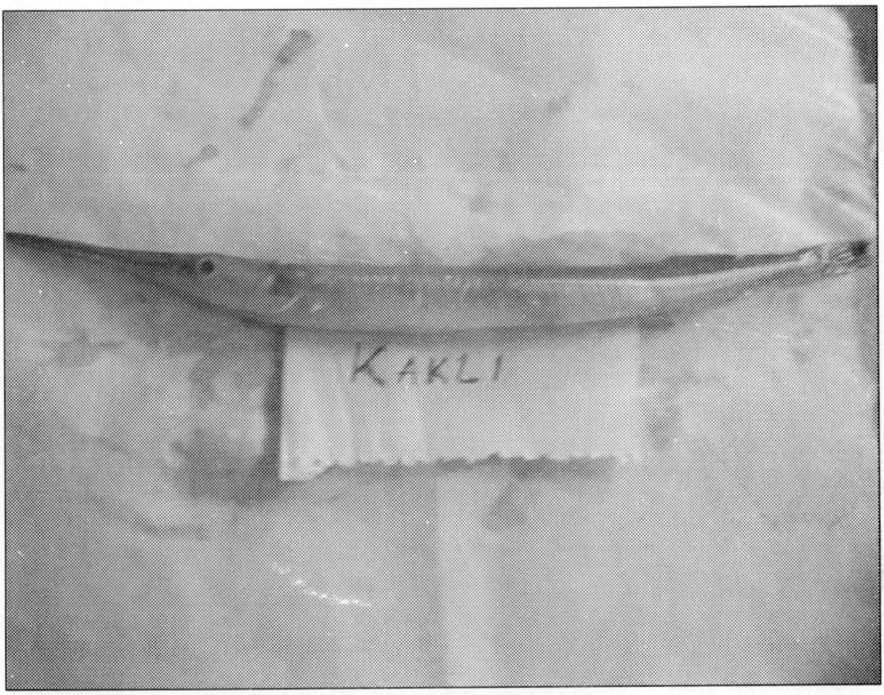

Plate 124: Kankli

Plate 125: Khoksa

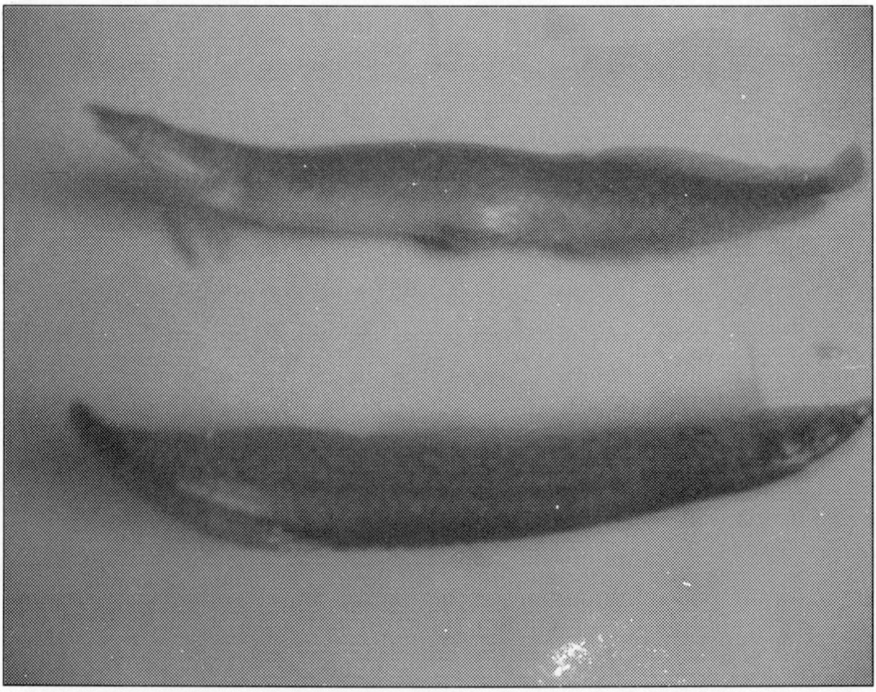

Plate 126: *Mastocembalus armetus* (Termas Tera)

Plate 127: *Macrognathas pankalus* (Pankal)

Plate 128: *Silurus cochinchinensis* (Pabda)

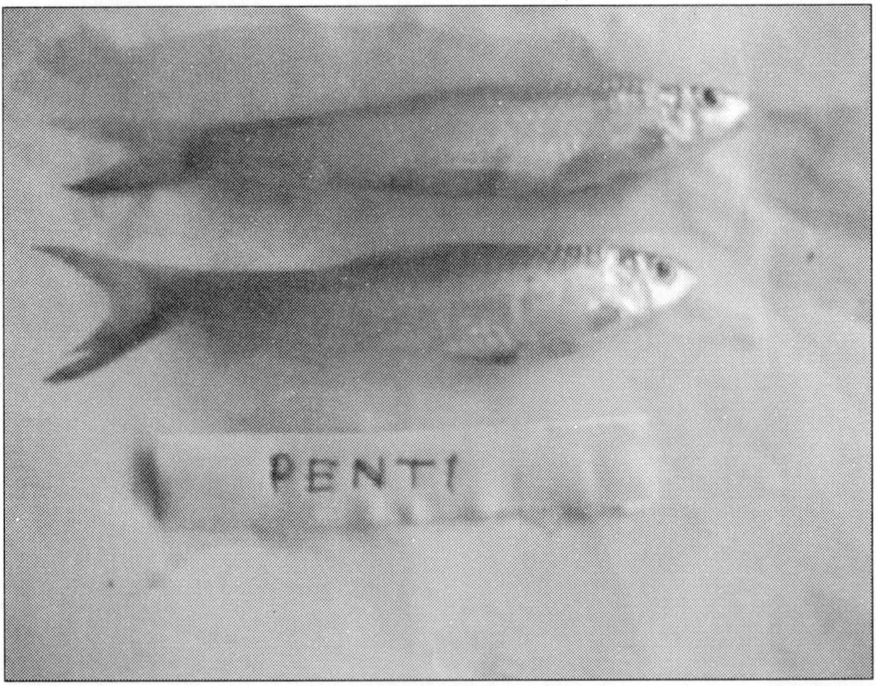

Plate 129: Penti

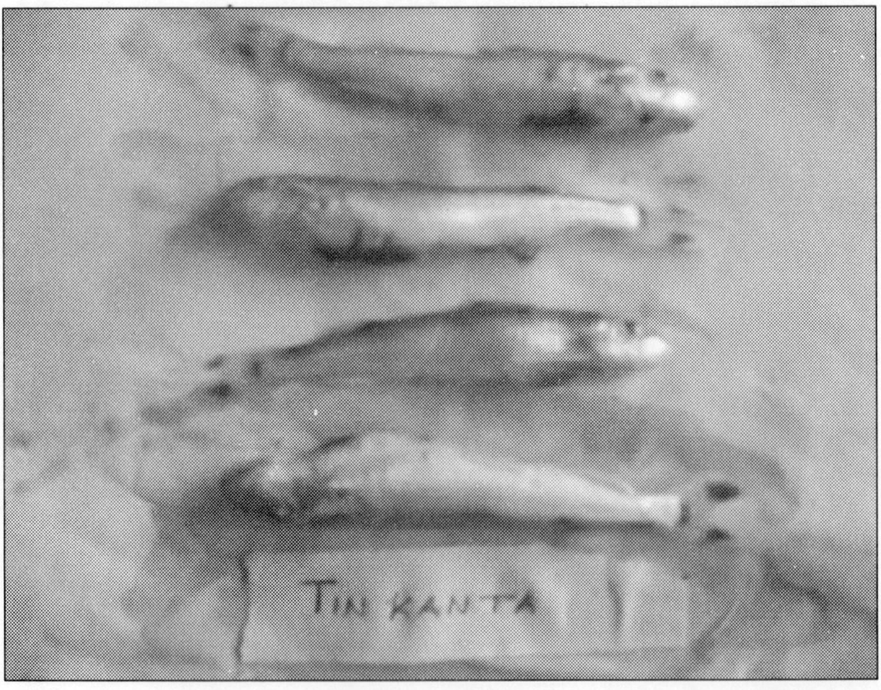

Plate 130: Tin kanta

Plates 131–132: Stone Cutter

Plate 131

Plate 132

Plate 133: Bulk of Napit Fish

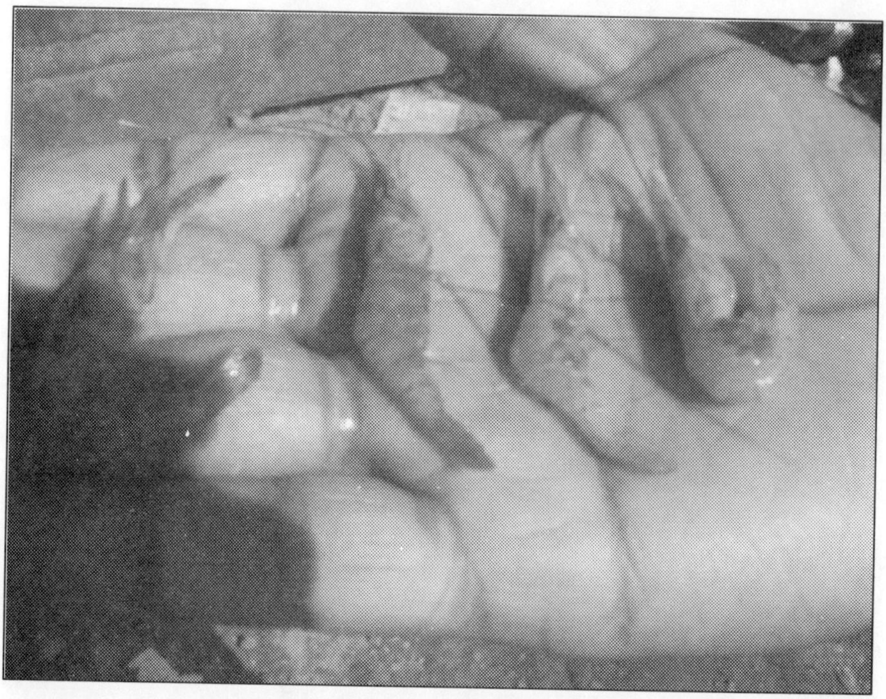

Plate 134: *Macrobrachium rudi*

Plates 135–138: *Macrobrachium assamensis* (*Jhinga maach*)

Plate 135

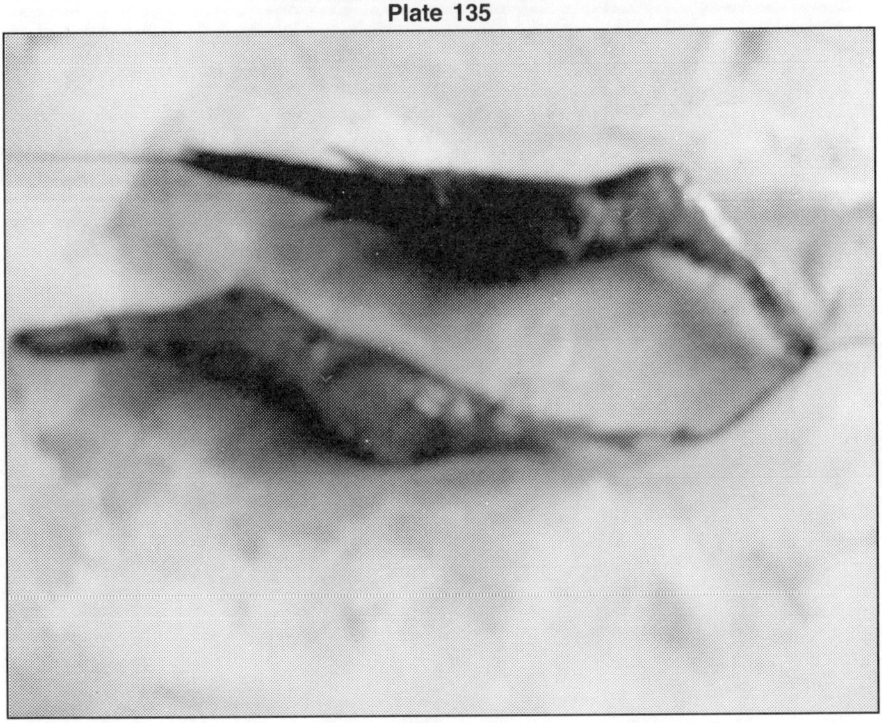

Plate 136

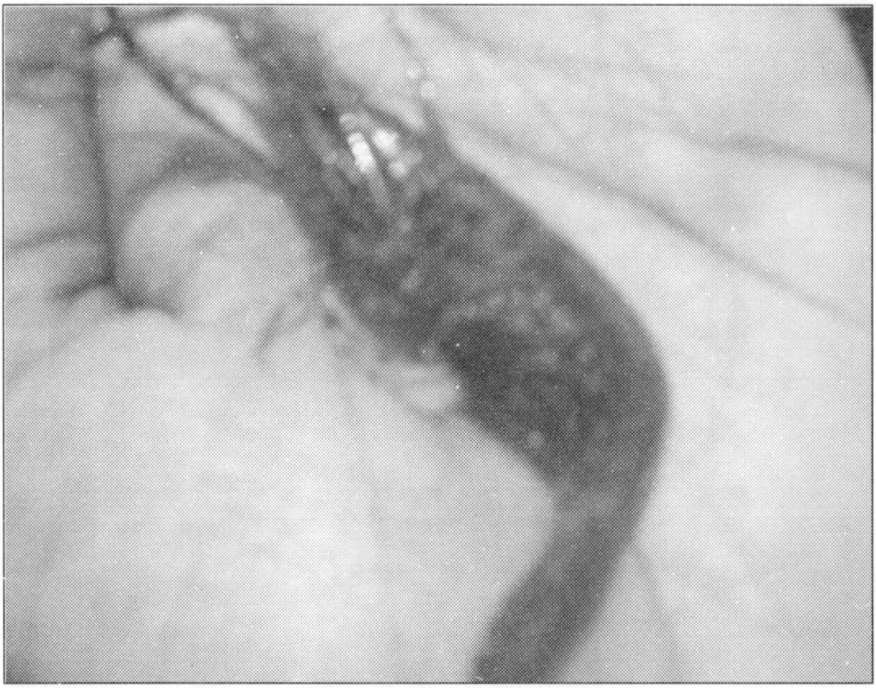

Plate 137

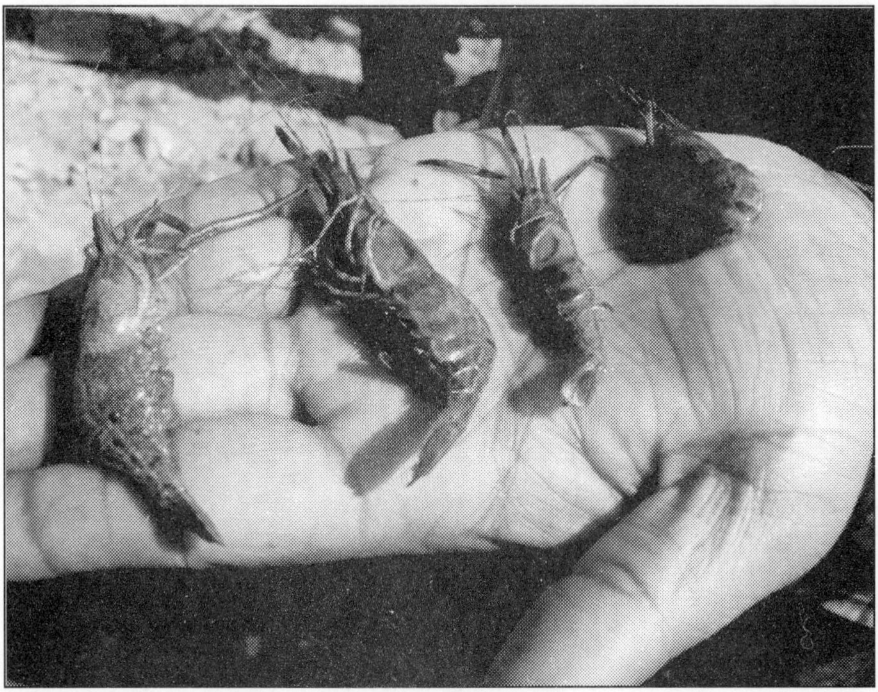

Plate 138

Chapter 5

Ecosytem Approach to Aquaculture

5.1. Topographical and Biological Conditions of Coldwater Area (Darjeeling District)

Topographical and biological conditions of the hill streams of Darjeeling hill areas are briefed (Sen and Biswas, 2006):

1. *Maximum and Minimum temperature of air and water in the area*:

   Maximum: Air : 32°C; Water: 29°C

   Minimum: Air : 17°C; Water: 15°C
2. *Conditions of stream, nature of stream*: Torrential, flowing through deep gorges with steep wooded sides. Their beds are covered with boulders, rocks and stones over which water flow in series of water falls, cascades and rapid with pools of various sizes here and there.
3. *General topography of the streams i.e. it is snow fed or rain fed*: During summer it is snow fed and spring fed, while in monsoon the streams are heavily fed with rain water.
4. *Acidity and alkalinity of water*: pH range 7.1 to 8.4.
5. *Nature of insects fauna in the stream bed*: Earth worms, round worms, insect larvae, larvae of simuliidac mostly chironomids, smaller crustaceans, may fly larvae, mosquito larvae, corixid bugs, nymphs, pleoptera.
6. *The level to which water is reduced during droughts before monsoon*: A maximum depth retained before monsoon in small streams is 3 ft. and in large streams is 10 ft.
7. *Existence of plant life in the river and its nature*: Diatoms predominate as attached organisms. The desmids, closterium and cosmariam present.

5.2. Natural Breeding Ground of Indigenous Fish

The natural breeding ground of hill stream fishes ared depicted in Plates 139–144.

(*i*) Bele (*Glossogobius giuris*)

Balasun, Mahananda *i.e.* 10 km from Siliguri during the month of December.

(*ii*) Chanda (*Parambassis nama*)

Found in Sevoke 15 kms frm Siliguri. The amount of this species is very less.

(*iii*) Katli (*Neolissocheilus hexagonalepis*)

Found in Teesta, Gish, Relli, Reyang post monsoon (Sept to Feb). This fish is found mostly near human settlements. Fish seeds are found in the clear waters. It has been observed that this species found in almost all rivers and tributaries of Darjeeling Dist. Almost 60-70 per cent of the total existing cold-water species is dominated by this species. It is also observed that the species is very hardy and thrives almost on any diet available in the surrounding. Preferably insects and any other slaughter waste are the favourite food. Grass like halhallia, Napier and any other soft grasses also preferred. It is also one of the best sports fish beside Mahseer and trout. This species is most preferable for both culture as well as table fish. The price of the fish during season time *i.e.* during Sept-Oct., when it migrates down stream is highest.

(*iv*) Chepti (*Cyprinion semiplotum*)

Found mostly in river Rishi only. It is to be found almost in all the rivers of Darjeeling district at one time but due to various natural and man made environmental stress the fish is near extinct. This fish is a delicacy and is a most priced ornamental fish species among other fish species of the hills. The size of the fish ranges from 250-500gms. The main identifying character of this species is the yellow stripes running across its fins.

(*v*) Kholsa (*Polyacantbus fasciata*)

This species is mostly found in the lower stretches river Teesta Relli, Balasun and Mahananda. It is also observed at Mirik Lake. It is also an ornamental fish species. Size of this hill Kholsia is slightly smaller and brightly coloured.

(*vi*) Pahari Pabda (*Silurus cochinchinensis*)

This is a unique find which is found in the Tirsat Khola during March in the clear and clean water near Kali Jhora. Water temperature suitable is ranging from 15-20 degree celcius.

(*vii*) Asala-naak Katwa (*Schizothorax Richardson, Schizothorax progastus*)

This fish dominates the rivers of Teesta, Rangeet and Gish. The seedlings are found in most of the tributaries of Teesta during September to March. Peak period is in the month of March around the confluence of river Teesta and Rangeet, Relli and Teesta, Teesta and Tar Khola.

Plates 139–140: Natural Nursery Ground at Reyang River

Plate 139

Plate 140

Plate 141: Human Settlement at River Teesta–One of the Main Reasons for Displacement of Endangered Coldwater Fish Species

Plate 142: Balasun River Site–Impact of Siltation Causing Destruction of Breeding Grounds

Plate 143: Rampant Use of Pesticides in Tea Gardens

Plate 144: Discharge of Effluents from Cinchona Factory Mongpu, Near Kalijhora

(viii) Kapre/Stone Cutter (*Pseudocheneius sulcatus*)

Found in River Teesta and Rishi. This fish is very hard to catch from the wild and has a good potential as an Aquarium fish. There are two species available one with, Sucker disc at the ventral side of the body for holding on to the rock and one without the sucker. Both the species are available in good number and it is also very delicacy.

(ix) Paheli Sar or Maha Sol (*Tor putitora*)

This fish is found in the river Teesta. But the fingerlings of up to 5 inch are only to be found in the rivers like Reyang, Relli, Rishi during post monsoon. It has been observed that the fish breeds during the heavy rainfall in hill areas when river Teesta in full flow with high turbid water and clear water coming from streams at different points of river Teesta.

5.3. Water Quality Analysis

Five sites (Tables 52–56) were selected for various water quality analyses as shown below. The sites are Sevok, Reyang, Teesta Bazar, Rangeet and Melli covering the entire stretch of river Teesta and its tributaries in the Darjeeling hill areas. Water samples (Plates 145–166) were collected from these sites fortnightly at the morning hours. The water pH found 8.04-8.54 and DO range was from 12.13-12.66. The range of average water temperature was 17.06-19.38.

Table 52: Water Quality Analysis Report at Site 1–Sevoke

Sampling Spot No.	*pH*	*Conductivity (mohs)*	*Temp°C*		*D.O. (mg/l)*	*Free CO_2 (mg/l)*	*Total Alk (mg/l)*	*Hard-ness (mg/l)*	*Chlorides (mg/l)*
			A	*W*					
1	8.1	105	28	20	12	11	10.2	30	4.26
2	8.4	98	25	20	12	10.12	10	31	4.21
3	8.5	84	24	19	12.3	10.56	10	29	4.5
4	8.4	105	24	17	12.5	10.8	10.1	28	4.3
5	8.2	78	24	17	12.5	10.7	10.1	29	4.1
6	8.2	77	23	16	12.6	10.6	10.2	30	4.2
7	8.2	80	21	16	13.2	10.5	10	31	4.3
8	8.2	81	22	16	13	10	10.1	30	4.2
9	8.6	80	22	17	13.2	10.1	10.1	31	4.1
10	8.4	80	23	16	13	10.2	10.2	30	4.3
11	8.3	90	31	19	12.5	10	10.3	30	4.1
12	8.4	91	29	20	12.4	10	10.2	32	4
13	8.2	85	28	18	12.8	10.6	10	28	4.3
14	8.3	89	30	19	12.6	10.7	10.1	29	4.1
15	8.1	85	30	19	12.9	9.9	10.2	29	4.2
16	8	105	32	20	13	10	10	31	4.3
Average	8.28	88.31	26.00	18.06	12.66	10.36	10.11	29.88	4.22

Table 53: Water Quality Analysis Report at Site 2–Reyang

Sampling Spot No.	pH	Conductivity (mohs)	Temp°C A	Temp°C W	D.O. (mg/l)	Free CO_2 (mg/l)	Total Alk (mg/l)	Hard-ness (mg/l)	Chlorides (mg/l)
1	8.4	70	27	22	10.8	8.5	20.5	30.1	4.5
2	8.3	68	27	22	10.6	8.4	22	30.2	4.3
3	8.1	52	26	22	10.2	8.5	20.4	30.5	4.6
4	8.2	48	26	22	10.6	8.3	20.2	30.3	4.5
5	8	62	25	20	10.6	8.3	20.3	30.6	4.3
6	8.2	63	25	19	10.5	8.2	20.2	30.1	4.2
7	8.2	58	18	16	10.4	7.9	30	30.2	4.2
8	7.8	53	19	16	10	7.8	25	30.3	4.1
9	7.8	80	19	16	13.2	7.7	28	30	4.3
10	8	81	19	15	13	7.9	29	30.1	4.4
11	8.2	87	30	19	12.8	7.9	29	30.5	4.5
12	8	84	26	20	12.5	8	28	30	4.6
13	8.2	80	28	21	12.8	8.4	21	32	4.2
14	8.2	81	27	19	12.6	8.3	22	32.5	4.4
15	8	53	29	20	12.9	7	20.5	30.1	4.5
16	8.2	62	28	21	12.5	6.9	21.5	30.3	4.2
Average	8.11	67.63	24.94	19.38	11.63	8.00	23.60	30.49	4.36

Table 54: Water Quality Analysis Report at Site 3–Teesta Bazar

Sampling Spot No.	pH	Conductivity (mohs)	Temp°C A	Temp°C W	D.O. (mg/l)	Free CO_2 (mg/l)	Total Alk (mg/l)	Hard-ness (mg/l)	Chlorides (mg/l)
1	7.9	52	23	17	12.5	6.6	20.4	29	4.2
2	7.5	50	27	17	12.3	6.5	22.2	28	4.3
3	7.2	70	22	17	12.2	6.6	20.6	30	4.7
4	8.1	75	22	16	12.1	6.7	20.3	32	4.6
5	8.2	78	20	16	12.3	6.6	20.5	36	4.8
6	8.3	78	20	17	12.1	6.5	20.4	38	4.9
7	8.5	72	17	15	12.7	6.19	28.5	29	5.1
8	8.4	73	17	14	12.8	7.8	29.6	29	5
9	8.2	80	18	16	13.1	7.3	30	30	4.7
10	8.3	81	19	18	13	7.2	31	31	4.3
11	8.3	72	25	19	12	7.1	35	32	4.5
12	8.2	69	26	19	12.2	7.2	25	30	4.3
13	8	70	26	18	12.3	6.6	28	28	4.2
14	8.1	68	27	17	12.1	6.7	28.5	29	4.4
15	7.4	75	27	19	12.3	5.9	29	25	4
16	8	69	27	18	12.1	5.8	26.5	28	4.1
Average	8.04	70.75	22.69	17.06	12.38	6.71	25.97	30.25	4.51

Table 55: Water Quality Analysis Report at Site 4–Rangeet

Sampling Spot No.	pH	Conductivity (mohs)	Temp°C		D.O. (mg/l)	Free CO_2 (mg/l)	Total Alk (mg/l)	Hard-ness (mg/l)	Chlorides (mg/l)
			A	W					
1	8.5	85	24	19	12.1	6.6	40	22	5.2
2	8.6	90	23	18	12.2	6.5	45	23	5.1
3	8.6	95	22	18	12.5	6.2	48	25	5.1
4	8.4	90	21	17	12.6	6.1	47	26	5
5	8.7	90	21	15	12.1	6.5	48	30	5.11
6	8.4	62	21	15	12.1	6.5	45	32	5.1
7	8.8	78	17	15	13	6.8	42	32	4.2
8	8.7	53	17	16	13.1	6.7	43	31	4.1
9	8.8	88	19	15	10.5	6.7	47	34	4.1
10	8.7	87	18	16	10.8	6.1	46	36	4.3
11	8.8	71	31	20	10.5	6.2	45	31	4.1
12	8.6	72	30	19	12.2	7.5	45	26	4.2
13	8.5	75	31	18	12.3	6.8	42	28	4.8
14	8.4	73	30	17	12.5	6.6	43	26	4.9
15	8.2	83	26	18	12.9	6	40	25	4.2
16	8	80	28	19	12.6	5.8	41	27	4.6
Average	8.54	79.50	23.69	17.19	12.13	6.48	44.19	28.38	4.63

Table 56: Water Quality Analysis Report at Site 5–Melli

Sampling Spot No.	pH	Conductivity (mohs)	Temp°C		D.O. (mg/l)	Free CO_2 (mg/l)	Total Alk (mg/l)	Hard-ness (mg/l)	Chlorides (mg/l)
			A	W					
1	8.1	72	25	17	13	6.6	30.5	38.5	4.5
2	8.2	78	25	16	12.8	6.9	31.6	36.5	4.3
3	8.1	72	24	16	13	7.1	32.5	37.5	4.5
4	8	73	24	15	13.2	7.2	33.6	38.5	4.6
5	8.2	70	22	15	13.6	7.3	39.1	37	4.7
6	8.2	71	21	15	13.1	7.5	38.2	36	4.3
7	7.5	69	16	14	13.4	7.5	35.1	38	4.2
8	7.8	68	17	14	13.3	7.6	36.1	39	4.2
9	8	77	24	15	9.4	7.6	33.1	32.1	4.1
10	8.1	79	23	14	10.5	7.6	32.1	38.1	4
11	8.1	76	29	19	12.5	7.6	30.1	39.1	4.1
12	8	70	28	18	12.6	7.3	30	38	4
13	8.2	72	29	19	12.9	7.2	35	38	4
14	8.2	75	27	18	13	7.1	33	36	4.3
15	8	76	27	19	13.5	6.5	30	32.2	4.2
16	8.1	79	29	20	13.2	6.2	31	33	4.3
Average	8.05	73.56	24.38	16.50	12.69	7.18	33.19	36.72	4.27

Plates 145–146: Water Analysis being done in CWF Laboratory at Reyang Fish Seed Farm

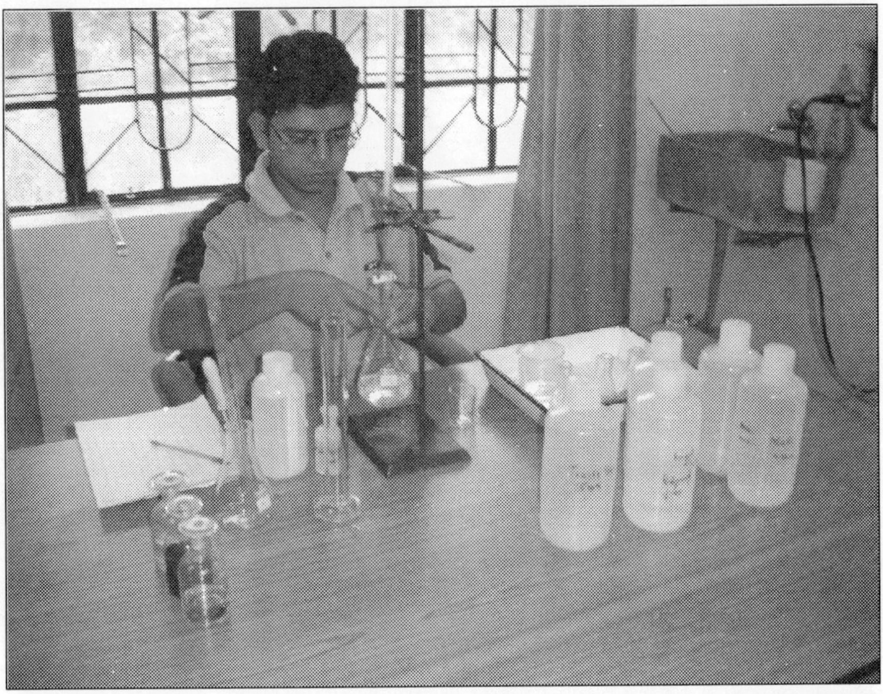

Plate 145

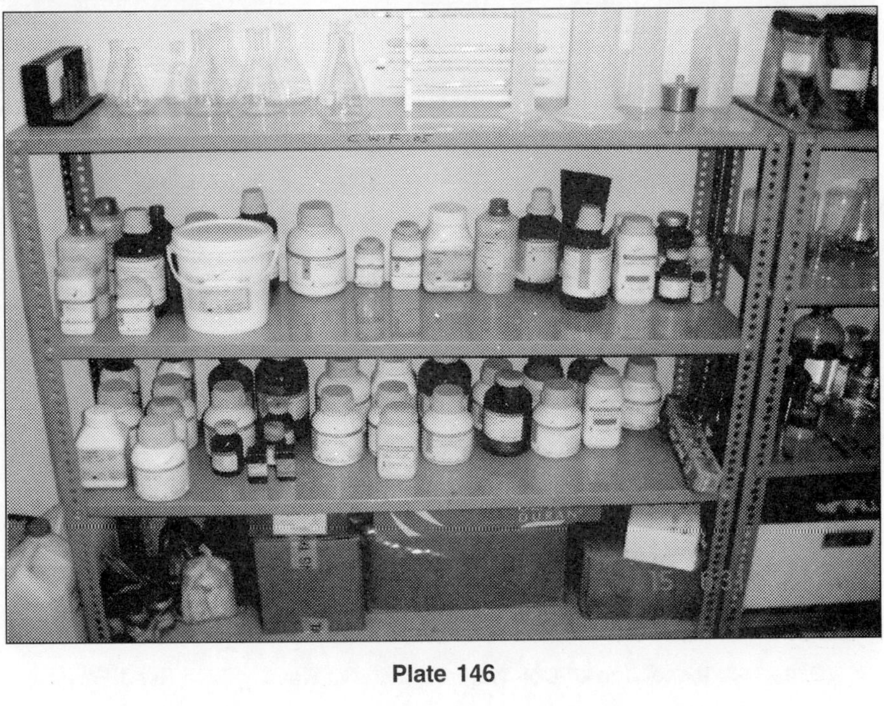

Plate 146

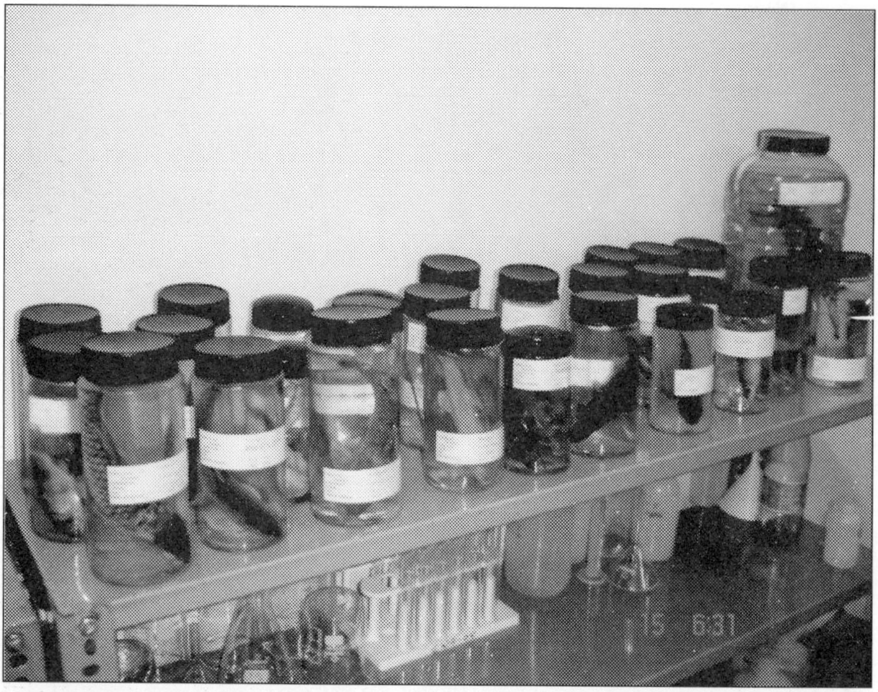

Plate 147: Fish Samples Collected in CWF Laboratory at Reyang Fish Seed Farm

Plate 148: Inspection of Coldwater Species at Reyang Fish Seed Farm

Plates 149–150: Computation of Data and Fish Identification in CWF Laboratory by Research Scholars

Plate 149

Plate 150

Plates 151–158: Water Samples being Collected at Various Sites of the River Teesta

Plate 151

Plate 152

Plate 153

Plate 154

Plate 155

Plate 156

Plate 157

Plate 158

Plates 159–160: Collection of Fingerlings at Reilli

Plate 159

Plate 160

Plate 161: Oxygenation of Jhora Pond

Plate 162: Sampling Site at Rangeet

Plate 163: Sampling Site at Reyang

Plate 164: Sampling Site at Sevoke

Plate 165: Sampling Site at Teesta

Plate 166: Sampling Site at Melli

5.4. Physico-chemical Parameters of Different Rivers of North Bengal

1. The physiochemical parameters of water (Tables 57–58) give an in sight on the extent of pollution and nutrient level. Therefore during the present study about North Bengal indigenous fish species, the estimation of routine parameters of water samples were done to determine the pollution stress at different regions at the study site.
2. Temperature is an important factor effecting the distribution of organism as well as the dissolved gasses in the water. It has been observed that with decrease in temperature the amount of DO is increased. In the present study the range of temperature was average 21°C at different station during monsoon and after monsoon period. The difference of temperature was due to seasonally, extent of sunlight, cloud cover, and altitude. pH shows the degree of hydrogen ion concentration in the water. This is extremely important factors which influences the various nutrients and ionic species in the water. The range of pH in water samples was 7.2 to 7.8. So we can say that the water is acidic, because of the acidic bottom soil.

Table 57: Physical Observation

Rivers	*Temp (Air)*	*Temp (Water)*	*Turbidity*	*Colour*
Teesta	25 °C	21°C	Slightly turbid	Slightly greenish brown
Mahananda	26°C	22 °C	Slightly turbid	–do–
Jaidhaka	26°C	21.5 °C	Slightly turbid	–do–
Balasun	25 °C	21.5 °C	Slightly turbid	Slightly brown
Buri Balasun	26 °C	22 °C	Slightly turbid	Slightly Turbid

Table 58: Chemical Observation

Rivers	*pH*	*DO) (mg/l)*	*Free CO_2 (mg/l)*	*Nitrate (mg/l)*	*Nitrite (mg/l)*	*Phosphate (mg/l)*
Teesta	7.8	5.6	5.5	0.25	0.075	0.45
Mahananda	7.5	5.7	4.8	0.28	0.075	0.50
Jaidhaka	7.5	5.4	4.5	0.3	0.070	0.60
Balasun	7.2	5.6	4.5	0.20	0.050	0.20
Buri Balasun	7.5	5.2	5.1	0,25	0.075	0.40

3. Dissolved oxygen is an important aquatic factor which decides the distribution of aquatic biota. The present study displays an average range 5.6 ppm. This is quiet satisfactory rather quiet high range. The lowest oxygen was recorded at Buri Balasun River where the water is quite polluted and water level is very low.

Table 59: Soil Analysis Report of Different Streams/Rivers in North Bengal

River's Name	*pH*	*Organic Carbon*	*Available Phosphate*	*Available Nitrogen*
Teesta	6.5	0.65 per cent	19 mg/100 g soil	2 mg/100 soil
Mahananda	6.2	0.70 per cent	20 mg/100 g soil	2.2 mg/100 g soil
Jaldhaka	6.3	0.60 per cent	23 mg/100g soil	2.5 mg/100 g soil
Balasun	5.5	0.55 per cent	17 mg/100 g soil	1.5 mg/100 g soil
Buri Balasun	5.1	0.59 per cent	20 mg/l00gsoii	2.2 mg/100 g soil

4. Phosphate is one of the most important factors in waters. But if the system is polluted by organic matter then it will be phosphate rich and will form algal bloom resulting eutrophication. Where as if phosphate content is low then we can say that nutrient content or chlorophyll concentration is low. The range of phosphate in present study was average 0.45 ppm. The lowest range was recorded from Balasun River (0.20 ppm).
5. Nitrogen is also an important factor shows the nutrient concentration in the water. The range of nitrogen in Nitrate form was 0.25–0.35 ppm. It shows that the Nutrient level of these waters is quiet low/normal.
6. Turbidity of the selected areas water samples is very high due to Heavy rain fall during monsoon month.
7. The free carbon di oxide is also an important factor shows the pollution level of waters. The range of nitrogen in Nitrate form was 0.25–0.35 ppm. The range of nitrogen in Nitrate form was 0.25–0.35 ppm. The range of CO_2 was 4.5-5.5 ppm, which is in normal range or moderately low, due to low organic carbon in river's bottom soil.
8. The out come of the current study can be visualized in terms of environmental quality and its influence on the fish diversity. The environment plays an important role in the disrupting the balance between the host and the pathogen. In many situation, the animal live a healthy normal life in the presence of pathogens, but when environmental stress factors increase, the tolerance tips in favors of the disease causing pathogens.The stress factors would normally affect the normal functioning such as breeding, feeding habit

In nutshell, Water Quality Parameters as measured in different rivers in Jalpaiguri District:

- ☆ Water temperature: 22 – 26°C
- ☆ pH: 6.9 – 7.4
- ☆ Total hardness: 40 –55mg/lt
- ☆ Ammonium (NH_4^+): 0.1 – 0.3mg/lt

- ☆ Nitrate (NO_3^-): 70 – 95mg/lt
- ☆ Nitrite (NO_2^-): 0.070 – 0.090mg/lt

In nutshell, Water Quality Parameters as measured in different Hilly Rivers:

- ☆ Water temperature: 12–15°C
- ☆ pH: 6.8 – 7.5 (*viz.* Ghorghoria -7.0, Jorai-7.0, Gadadhar-7.5, Kaljani-7.5)
- ☆ Ammonium (NH_4^+): 0.1mg/lt
- ☆ Nitrate (NO_3^-): 0.078mg/lt
- ☆ Nitrite (NO_2^-): 82mg/lt
- ☆ Total hardness: 30-35mg/lt

5.5. Soil Nutrients Condition of North Bengal

Soil nutrients maps and soil fertility status reports of different blocks of all the studied districts have been collected from the Directorate of Agriculture, GOWB. These are furnished in Tables 60–102 and Figures 11–20. These reports will be helpful during implementation of any fisheries and aquaculture project in this area. The soil and water can be treated accordingly for getting a successful crop.

Table 60: Physiography and Agroclimatic Regionalization of Kharibari Block, Darjeeling District

Climatic Feature	*March to May*	*June to October*	*November to February*
Rainfall (mm)	396.6	2134.0	42.6
Air Temparature–Max. °C	32.3	31.3	26.0
Air Temparature–Mini. °C	20.5	24.5	12.8
Relative Humidity (per cent)	60.0	81.0	70.0
Dew fall	Low significant	Not significant	Very significant

Source: Directorate of Agriculture, GOWB.

Table 61: Distribution of Land According to Identified Landform of Kharibari Block, Darjeeling District

Land Form Type	*Symbols Used*	*Area Covered*	*General Slope*	*Common Land Treatment Status*
Valley	3	182.5	0-1 per cent	Poorly bunded
Upper Piedmont alluvial fan	4	512.5	2-5 per cent	Untreated
Lower piedmont alluvial fan	5	1077.5	1-3 per cent	Untreated
Flood plain	6	7665.8	0-1 per cent	Bunded
Active Flood Plain	7	2790.0	0-1 per cent	Bunded to poorly bunded
Depressions	8	30.0	0-1 per cent	Poorly bunded
River and Homestead		2091.7		
Total		14350.0		

Source: Directorate of Agriculture, GOWB.

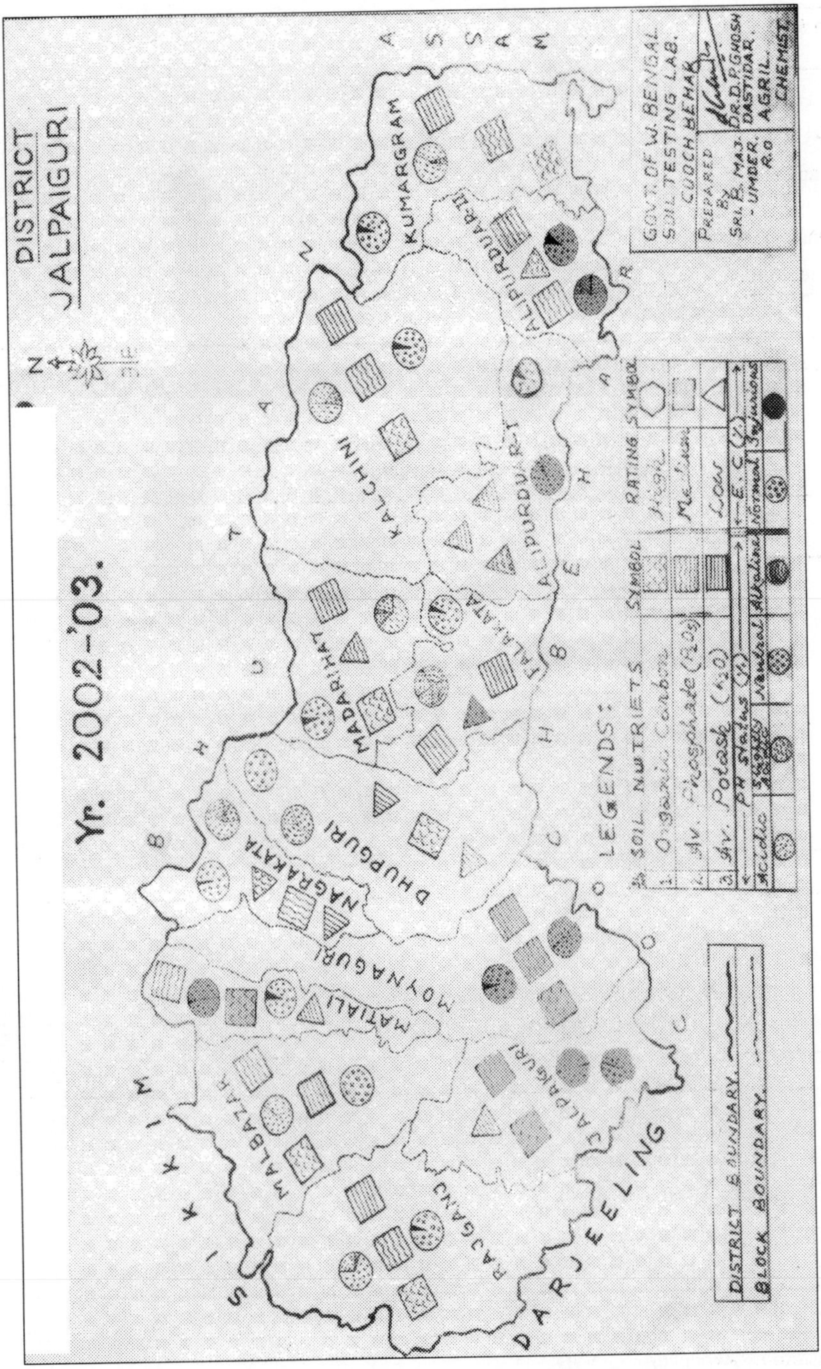

Source: Directorate of Agriculture, GOWB

Figure 11: Block-wise Soil Fertility Index Map of Jalpaiguri District

Table 62: Distribution of Area as per Present Landuse Pattern of Kharibari Block, Darjeeling District

Landuse Type	*Notations Used*	*Area (Ha.)*	*Remarks Specifying Landuse Feature*
Forest	F	575.0	Thick and dense
Tea Garden	Pt	1077.5	Dense T.G.
Cultivated Land	C	10605.8	Mostly single and some double and triple cropping
River and Jhora		710.0	
Homestead		1371.7	
Total		14350.0	

Source: Directorate of Agriculture, GOWB.

Table 63: Distribution of Soil as per Soil Series Identified of Kharibari Block, Darjeeling District

Notations	*Name of the Soil Series Identified*	*Area Coverage (ha)*	*Commonly Related Landform-Land Use Type*
S1	Tari, A-C, Sandy	2342.5	Valley-cultivated, Active flood plain-cultivated, Depression-cultivated, Valley forest
S2	Tukriajhar, A-B, Coarse Loamy	512.50	Upper piedment alluvial fanforest
S3	Kishoredoba, A1-A2-B2, Coarse Loamy	7665.80	Flood Plain cultivated
S4	Fulbari, A-B, sandy	1737.5	Lower Piedmont alluvial fan Tea Garden.
	Total	12258.0	

Source: Directorate of Agriculture, GOWB.

Table 64: Distribution of Soil According to Soil Reaction of Kharibari Block, Darjeeling District

Soil Reaction Type	*Area Distribution (ha.)*
Strongly Acidic	9886
Slightly Acidic-Neutral	2372.5
Slightly Alkaline-Neutral	Nil
Calcareous	Nil
Alkaline	Nil
Total	12258.5

Source: Directorate of Agriculture, GOWB.

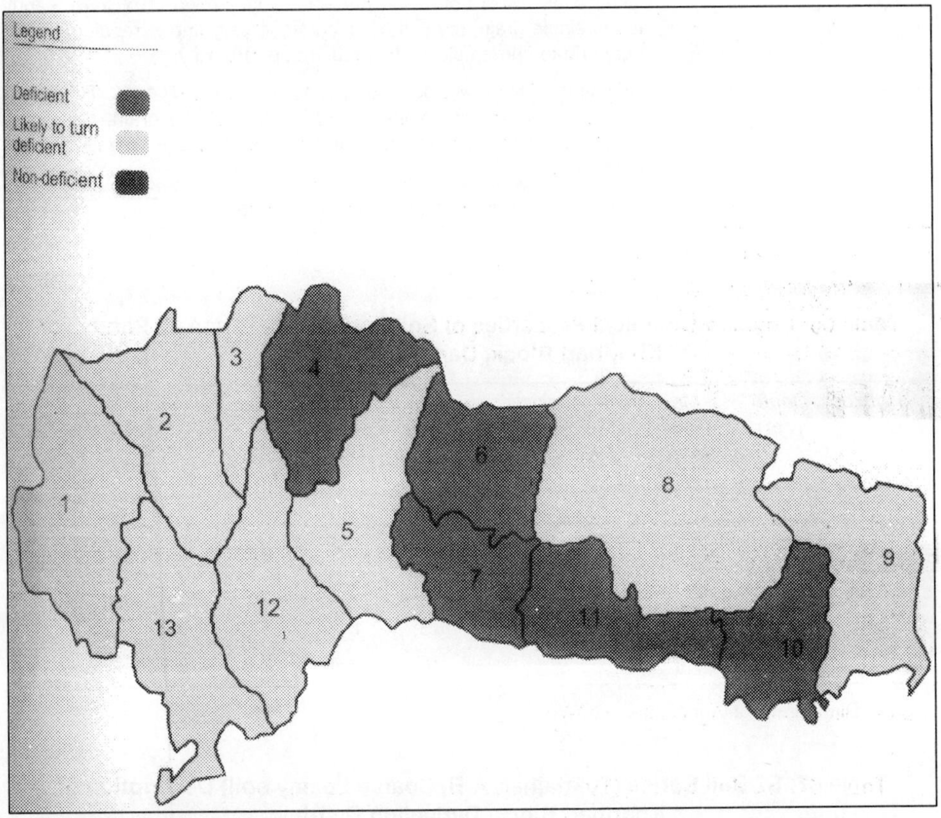

Source: Directorate of Agriculture, GOWB

Figure 12: Block-wise Status of Available Nitrogen of Jalpaiguri District 1: Rajganj; 2: Mal; 3: Metali; 4: Nagrakata; 5: Dhupguri; 6: Madarihat; 7: Falakata; 8: Kalchini; 9: Kumargram; 10: Alipurduar II; 11: Alipurduar I; 12: Moinaguri; 13: Jalpaiguri

Table 65: Soil Series (S1–Tari, A-C, Sandy Soil) Description of Kharibari Block, Darjeeling District

Horizon	Depth in cm	Description
Ap	0-11	Clear wavy lower boundary, dry soil colour light yellowish brown (19YR6/4) with many medium prominent red mottles, loamy sand, single grain, many very fine to medium pores, very fine to medium ample roots, rapid permeability.
Ap1	11-38	Clear wavy lower boundary, dry soil colour brown (19YR6/4), sandy loam, single grain, dry soft hard, common very fine to medium pores, very fine to fine ample roots, medium permeability.
Ap2	38-103	Diffuse irregular lower boundary, subsoil colour brown (10YR5/4) with common coarse faint mottles, loamy sand, single grain, dry loose, very fine to fine pores, very fine to fine profuse roots, medium permeability.
C	103+	Dry soil colour white, sandy, single grain, loose.

Source: Directorate of Agriculture, GOWB.

Table 66: Physico-Chemical Properties of Soil Series "S1–Tari, A-C, Sandy" of Kharibari Block, Darjeeling District

Horizon	Depth (cm)	Mechanical Analysis			Textural Class	Chemical Analysis			
		Sand (%)	Silt (%)	Clay (%)		pH	CEC Me/100 gm	EC Ms/cm	Organic Carbon (%)
Ap	0-11	69.4	23.4	7.2	Ls	5.9	3.6	0.08	1.09
A1.1	11-38	59.4	31.4	9.2	SL	5.6	4.8	0.19	1.17
A1.2	38-103	79.4	20.6		LS	5.9	3.2	0.04	1.73
C	103+	93.4	6.6		S	6.9	2.8	0.07	1.93

Source: Directorate of Agriculture, GOWB.

Table 67: S2 Soil Series (Tukriajhar, A-B, Coarse Loamy soil) Description of Kharibari Block, Darjeeling District

Horizon	Depth in cm	Description
$A_{1.1}$	0-10	Gradual irregular lower boundary, dry soil colour brown (7.5YR5/4), loamy, strong medium to coarse granular, many medium size coarse pores, medium to coarse size profuse roots, very slow permeability.
$A_{1.2}$	10-40	Gradual irregular lower boundary, dry sub soil colour yellowish red, loamy, strong medium granular, many medium to coarse pores, medium to coarse profuse roots, very slow permeability.
B	40+	Dry soil colour red (2.5YR4/6), Loamy, very strong medium granular, many medium pores, medium to coarse profuse roots, very slow permeability.

Source: Directorate of Agriculture, GOWB.

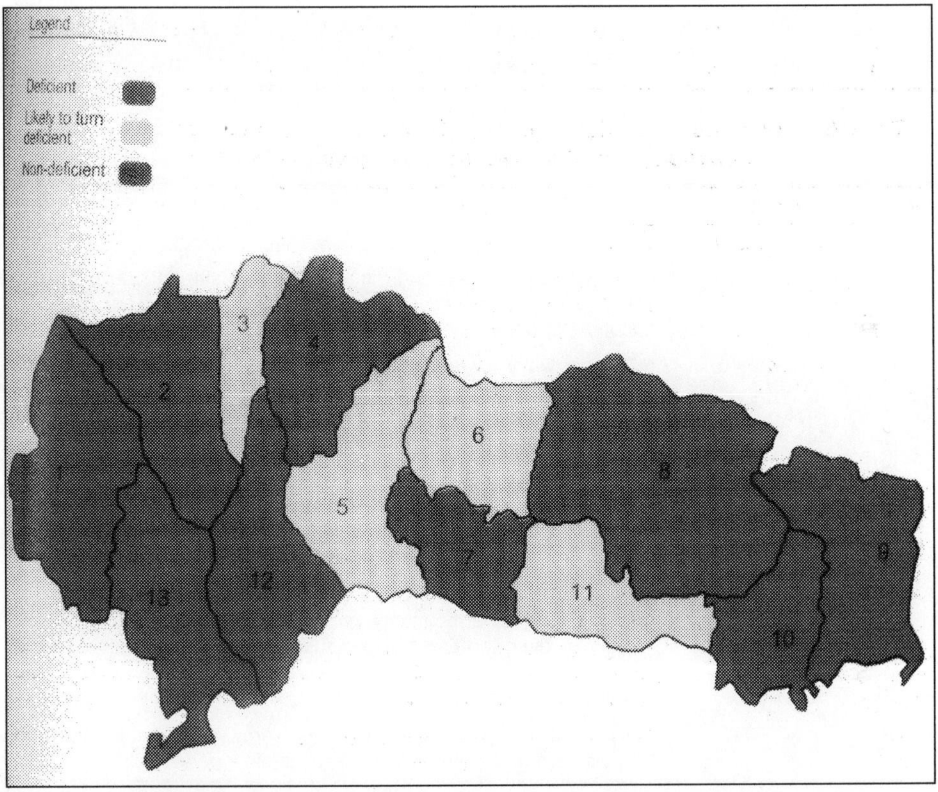

Source: Directorate of Agriculture, GOWB

Figure 13: Block-wise Status of Available Phosphorus of Jalpaiguri District
1: Rajganj; 2: Mal; 3: Metali; 4: Nagrakata; 5: Dhupguri; 6: Madarihat; 7: Falakata; 8: Kalchini; 9: Kumargram; 10: Alipurduar II; 11: Alipurduar I; 12: Moinaguri; 13: Jalpaiguri

Table 68: Physico-Chemical Properties of Soil Series "S2 Tukriajhar, A-B, Coarse Loamy" of Kharibari Block, Darjeeling District

Horizon	Depth (cm)	Mechanical Analysis			Textural Class	Chemical Analysis			
		Sand (%)	Silt (%)	Clay (%)		pH	CEC Me/100 gm	EC Ms/cm	Organic Carbon (%)
$A_{1.1}$	0-10	44.6	40.8	14.6	L	5.9	8.4	0.04	1.62
$A_{1.2}$	10-40	44.6	38.8	16.6	L	5.7	6.2	0.02	0.58
B	40+	35.2	36.2	28.6	L	5.6	4.8	0.02	1.09

Table 69: Soil Series (S3, Kishoredoba, A1-A2-B2, Coarse Loamy Soil Series) Description of Kharibari Block, Darjeeling District

Horizon	Depth in cm	Description
Ap	0-22	Gradual wavy lower boundary, dry soil colour pale brown(10YR6/3), loamy, moderately strong, fine to very fine sub-angular blocky, dry very hard, many very fine to medium pores, many very fine to medium roots, moderately slow permeability.
A_1	22-40	Gradual wavy lower boundary, moist soil colour brown (10YR4/3), loamy, moderately strong, fine to medium sub-angular blocky, moist very friable, many very fine to medium pores, many very fine to fine roots, moderately slow permeability.
A_2	40-64	Diffused wavy lower boundary, moist soil colour brown (10YR5/3), loamy with many coarse size diffused dark red mottles, very fine to fine single grain, moist very friable, very fine to fine many pores, thick continuous cutans, many fine to fine roots, moderately slow permeability.
$B_{2.1}$	64-77	Diffuse wavy lower boundary, moist soil colour brown (10YR5/3) with many medium to coarse size diffused yellowish red colour mottles, loamy, weak very fine to fine single grain, moist friable, many very fine to fine pores, very fine few roots.
$B_{2.2}$	77+	Dry soil colour light yellowish brown, loamy, single grain, moist very friable, very fine pores, very fine roots, slow permeability.

Table 70: Physico-Chemical Properties of Soil Series "S3, Kishoredoba, A1-A2-B2, Coarse Loamy" of Kharibari Block, Darjeeling District

Horizon	Depth (cm)	Mechanical Analysis			Textural Class	Chemical Analysis			
		Sand (%)	Silt (%)	Clay (%)		pH	CEC Me/100 gm	EC Ms/cm	Organic Carbon (%)
Ap	0-22	44.6	38.8	16.6	L	4.3	4.8	0.15	1.6
A_1	22-40	45.2	35.4	19.4	L	5.1	4.8	0.15	Traces
A_2	40-64	45.2	35.4	19.4	L	5.2	5.3	0.11	0.56
$B_{2.1}$	64-77	42.6	36.8	20.6	L	5.2	5.2	0.11	2.72

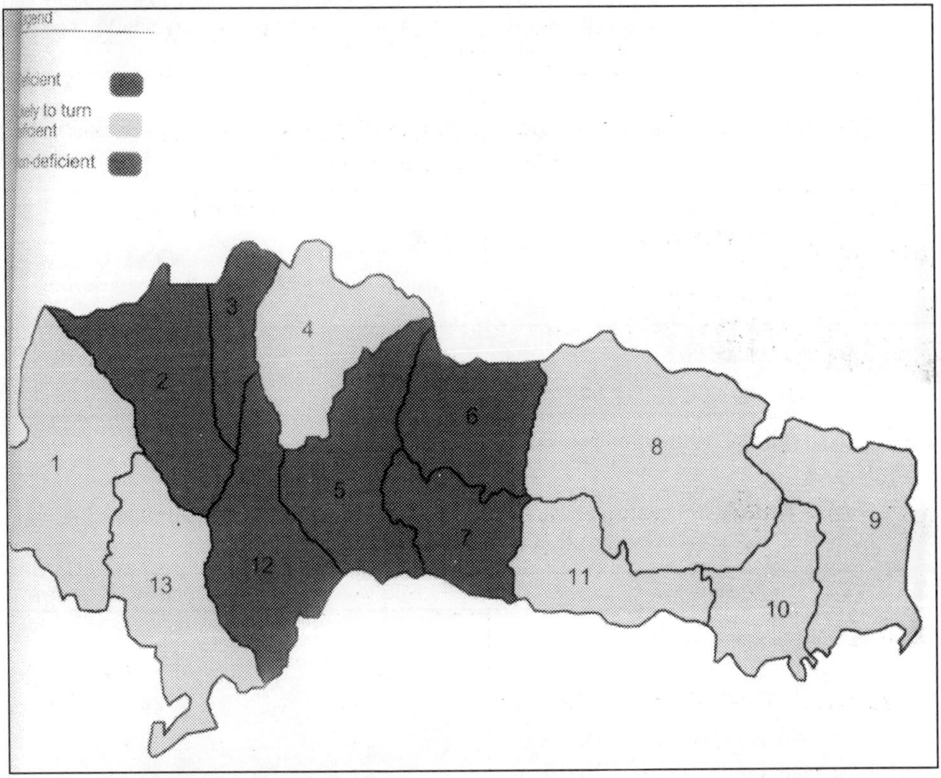

Source: Directorate of Agriculture, GOWB

Figure 14: Block-wise Status of Available Potassium of Jalpaiguri District
1: Rajganj; 2: Mal; 3: Metali; 4: Nagrakata; 5: Dhupguri; 6: Madarihat; 7: Falakata; 8: Kalchini; 9: Kumargram; 10: Alipurduar II; 11: Alipurduar I; 12: Moinaguri; 13: Jalpaiguri

Table 71: Soil Series (S4, Fulbari, A-B, Sandy) Description of Kharibari Block, Darjeeling District

Horizon	Depth in cm	Description
A_1	0-10	Gradual wavy boundary, dry soil colour light yellowish brown (10YR6/4), sandy loam, weak sub-angular blocky, dry soft hard, many fine to coarse pore, medium to coarse roots, moderately slow permeability.
B	10-75+	Dry soil colour strong brown (7.5YR5/6), sandy loam, strong sub-angular blocky, dry soft hard, common pores, very fine to fine roots, very slow permeability.

Table 72: Physico-Chemical Properties "S4, Fulbari, A-B, Sandy Soil Series" of Kharibari Block, Darjeeling District

Horizon	Depth (cm)	Mechanical Analysis			Textural Class	Chemical Analysis			
		Sand (%)	Silt (%)	Clay (%)		pH	CEC Me/100 gm	EC Ms/cm	Organic Carbon (%)
A_1	0-10	75.2	16.8	8.0	LS	4.8	4.6	0.08	0.72
B	10-75+	66.0	29.4	6.6	SL	5.3	5.8	0.04	0.66

Source: Directorate of Agriculture, GOWB.

Table 73: Physiography and Agroclimatic Regionalization of Naxalbari Block, Darjeeling District

Climatic Feature	March to May	June to October	November to February
Rainfall (mm)	376.6	2134.0	42.6
Air Temperature–Max. °C	32.3	31.3	26.0
Air Temperature–Mini. °C	20.5	24.5	12.8
Relative Humidity (per cent)	60.0	81.0	70.0
Dew fall	Low significant	Not significant	Very significant

Climatic condition of the block is of Tropical per humid type.

Source: Directorate of Agriculture, GOWB.

Table 74: Distribution of Land According to Identified Landform of Naxalbari Block, Darjeeling District

Land Form Type	Symbols Used	Area Covered	General Slope	Common Land Treatment Status
Back slope	2	20.00	3-5	Untreated
Valley	3	425.0	0-1	Poorly bunded
Piedmont fan (raised terraces mildly undulating)	5	9162.27	2-3	Untreated

Table 74–Contd...

Land Form Type	Symbols Used	Area Covered	General Slope	Common Land Treatment Status
Flood plain	6	3669.04	0-1	Bunded
Active flood plain	7	1327.50	0-1	Bunded
Misc. (River, Jhora, Home, Stead etc.)	9	3387.0		
Total		17990.81		

Source: Directorate of Agriculture, GOWB.

Table 75: Distribution of Area as per Present Landuse Pattern of Naxalbari Block, Darjeeling District

Landuse Type	Notations Used	Area (Ha.)	Remarks Specifying Landuse Feature
Tea Garden	Pt	5192.77	Dense T.G.
Forest	F	3142.50	Thick and denseforest of teak
Cultivated Land	C	5741.04	Mostly rainfed single to double crpped
Cultivated waste land	W1	452.50	Acquired by defence, remain untreated
Uncultivated	W2	75.00	Composed of stone and gravel, devoid of soil.
Non Agril. Use (River, Jhora, Homestead)		3387.3	
	Total	17990.81	

Source: Directorate of Agriculture, GOWB.

Table 76: Distribution of Soil as per Soil Series Identified

Notations	Name of the Soil Series Identified	Area Coverage (ha)	Commonly Related Landform-Land Use Type
S1	Chhoto Maniram, A-C, coarse loamy	1752.50	Valley-cultivated, Active flood plain
S2	Dalkajha,A-B-C, sandy	8544.77	Back slope-Forest, piedment alluvial fan forest and tea garden.
S3	Budhkarma, A1-A2-B2, coarse loamy	3669.04	Flood Plain cultivated
S4	Nipnia, A-B, sandy	637.5	Piedmont alluvial fan cultivated
	Total	14603.81	

Source: Directorate of Agriculture, GOWB.

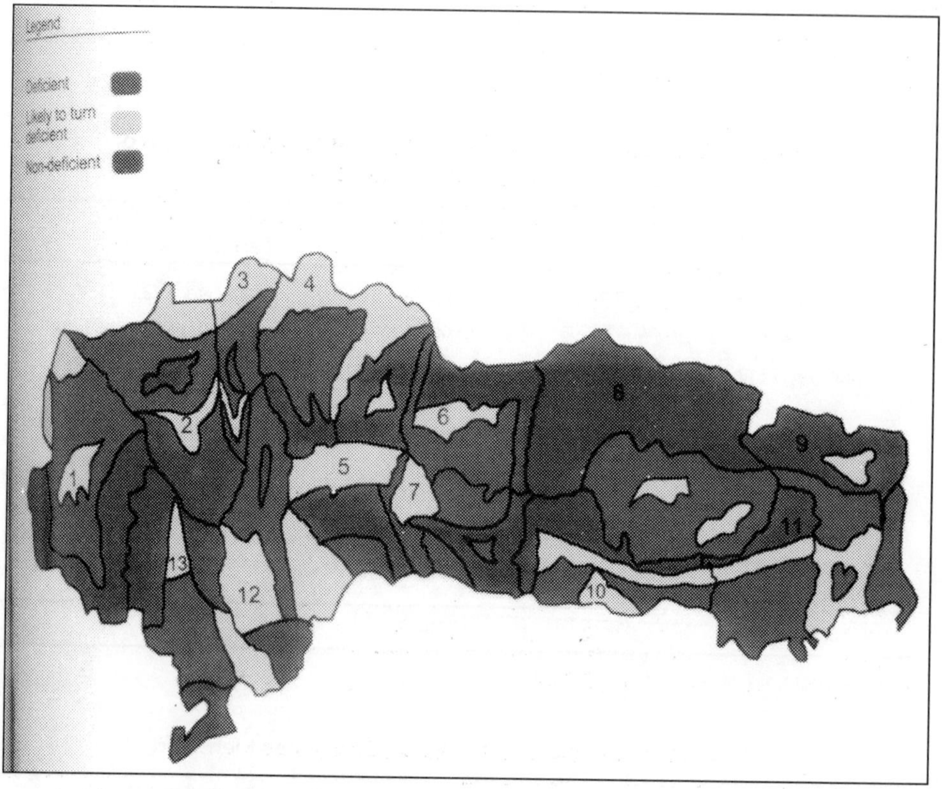

Source: Directorate of Agriculture, GOWB

Figure 15: Block-wise Status of Available Zinc of Jalpaiguri District
1: Rajganj; 2: Mal; 3: Metali; 4: Nagrakata; 5: Dhupguri; 6: Madarihat; 7: Falakata; 8: Kalchini; 9: Kumargram; 10: Alipurduar II; 11: Alipurduar I; 12: Moinaguri; 13: Jalpaiguri

Table 77: Distribution of Soil According to Soil Reaction of Naxalbari Block, Darjeeling District

Soil Reaction Type	Area Distribution (ha.)
Strongly Acidic	9182.27
Slightly Acidic-Neutral	5421.54
Slightly Alkaline-Neutral	–
Calcareous	–
Alkaline	–
Total	14603.81

Source: Directorate of Agriculture, GOWB.

Table 78: Soil Series Description of "S1–Chhoto Maniram, A-C, Coarse Loamy" of Naxalbari Block, Darjeeling District

Horizon	Depth in cm	Description
Ap	0-12	Clear irregular lower boundary, dry soil colour grayish brown (10YR5/2), with many medium to coarse diffused mottles of dark reddish grey colour(2.5YR4/1), loamy sandy texture, weak very fine to medium sub-angular blocky structure, dry hard, very fine to medium many pores, very fine to fine profuse root, moderate permeability.
A1.2	40-55	Clear irregular lower boundary, dry soil colour light brownish grey (10YR6/2), with common diffused reddish yellow colour mottles (7.5YR6/6), loamy texture with medium to coarse size profuse(>50 per cent),stone and gravells, massive structure, very fine to medium pores, very fine to fine roots, moderate permeability.
C	55+	Dry soil colour light brownish grey (10YR6/2), fine to medium diffused reddish yellow mottles (7.5YR6/6), sandy loam with medium to coarse size stone and gravel (>25 per cent), massive, few fine pores, few very fine to fine roots, moderately slow permeability.

Source: Directorate of Agriculture, GOWB.

Table 79: Physico-Chemical Properties of Soil Series "S1–Chhoto Maniram, A-C, Coarse Loamy" of Naxalbari Block, Darjeeling District

Horizon	Depth (cm)	Mechanical Analysis			Textural Class	Chemical Analysis			
		Sand (%)	Silt (%)	Clay (%)		pH	CEC Me/100 gm	EC Ms/cm	Organic Carbon (%)
A1.1	12-40	53.2	33.4	13.6	SL	5.4	5.9	0.27	0.17
A1.2	40-55	45.2	37.4	17.4	L	5.3	6.5	0.34	1.2
C1	55+	55.2	29.4	15.4	SL	5.4	5.8	0.41	0.78

Source: Directorate of Agriculture, GOWB.

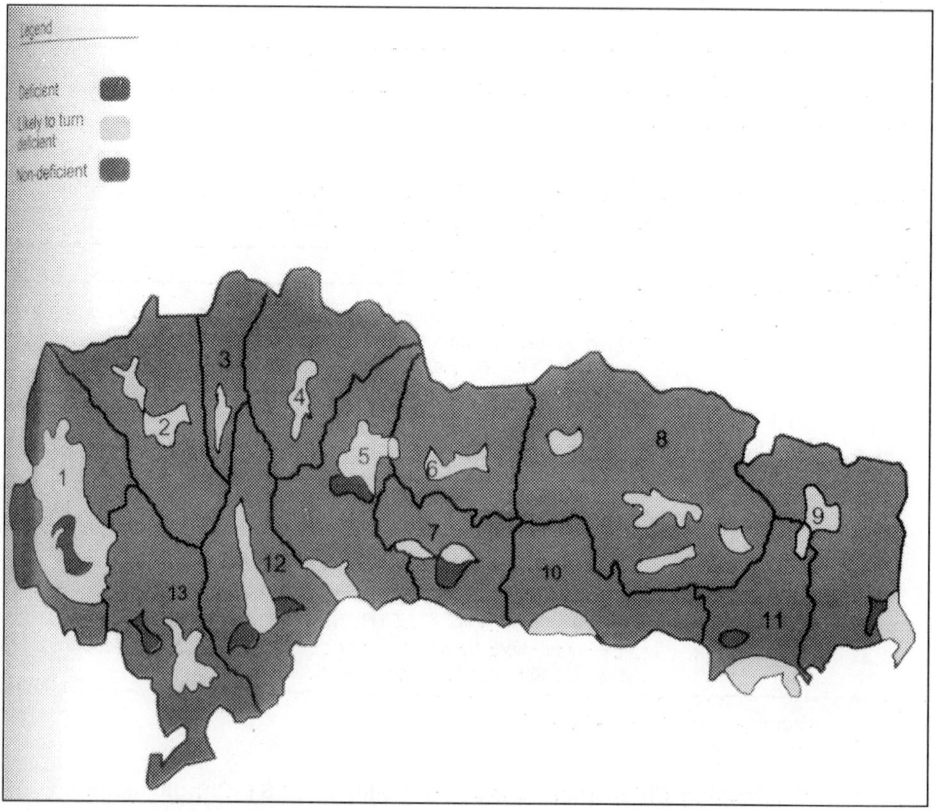

Source: Directorate of Agriculture, GOWB

Figure 16: Block-wise Status of Available Sulphur of Jalpaiguri District
1: Rajganj; 2: Mal; 3: Metali; 4: Nagrakata; 5: Dhupguri; 6: Madarihat; 7: Falakata; 8: Kalchini; 9: Kumargram; 10: Alipurduar II; 11: Alipurduar I; 12: Moinaguri; 13: Jalpaiguri

Table 80: S2: Soil Series (Dalkajha, A-B-C, sandy) Description of Naxalbari Block, Darjeeling District

Horizon	*Depth in cm*	*Description*
$A_{1.1}$	0-20	Clear wavy lower boundary, dry soil colour yellow (10YR7/6), loamy sand, medium to coarse stone and gravels (50 per cent), moderately strong, medium sized sub-angular blocky, dry very hard, many very fine to coarse pores, many very fine to coarse roots, medium permeability.
$A_{1.2}$	20-34	Gradual, wavy lower boundary, dry soil colour light yellowish brown (10YR6/4), loamy sand with medium to coarse size stone and gravel (50 per cent), single grain, loose, many very fine to medium pores, very fine to coarse ample roots, moderately rapid permeability.
AB	34-48	Clear, wavy lower boundary, dry soil colour brownish yellow (10YR6/6), loamy sand with medium to coarse stone and gravel(50 per cent), single grain, loose, very fine to medium many pores, very fine to coarse profuse roots, moderate permeability.
Cambic B	48-84	Clear wavy lower boundary, dry soil colour yellow, sandy loam with stone and gravel(<25 per cent), weak, very fine to medium sub-angular blocky, dry soft hard, common medium pores, very fine few roots, moderately slow permeability.
C	84+	Dry soil colour, very pale brown, loamy, sand with ample stone and gravel (>90 per cent).

Table 81: Physico-Chemical Properties of Soil Series "S2: Dalkajha, A-B-C, Sandy" of Naxalbari Block, Darjeeling District

Horizon	*Depth (cm)*	*Mechanical Analysis*			*Textural Class*	*Chemical Analysis*			
		Sand (%)	*Silt (%)*	*Clay (%)*		*pH*	*CEC Me/100 gm*	*EC Ms/cm*	*Organic Carbon (%)*
$A_{1.1}$	0-20	84.6	2.2	13.2	LS	5.1	4.3	0.08	0.66
$A_{1.2}$	20-34	84.6	6.2	9.2	LS	5.2	4.5	0.03	1.46
AB	34-48	75.2	16.8	8.0	LS	4.9	4.6	0.08	0.72
B	48-84	58.6	27.4	14.0	SL	5.3	8.7	0.03	0.49
C	84+	84.6	9.4	6.0	LS	6.1	2.4	0.02	0.77

Source: Directorate of Agriculture, GOWB.

Table 82: S3: Soil Series (Budhkarma, A1-A2-B2, Coarse Loamy) Description of Naxalbari Block, Darjeeling District

Horizon	*Depth in cm*	*Description*
Ap	0-22	Gradual wavy lower boundary, dry soil colour light brownish grey(10YR6/2), with many very fine to medium diffused strong brown mottles, loamy, very fine to medium strong, sub-angular blocky, dry excessive hard, many very fine to coarse pores with long cracks, many very fine to fine ample roots, medium permeability.

Contd...

Table 82–Contd...

Horizon	*Depth in cm*	*Description*
A_1	22-38	Gradual, wavy lower boundary, dry soil colour light brownish grey (10YR6/2), loamy, very fine to fine strong sub-angular blocky, very hard, many very fine to coarse pores, very fine to fine ample roots, medium permeability.
A_2	38-111	Gradual, irregular lower boundary, dry soil colour, very pale brown (10YR7/4) with few very fine to medium diffused mottles, colour yellowish brown, sandy loam, weak very fine to medium subangular blocky, dry hard, very fine to medium many pores, thick continuously clay cutans, very fine to fine common roots, medium permeability.
B_2	111+	Dry soil colour light yellowish brown, sandy loam, very fine moderately strong, sub-angular blocky, dry hard, many very fine to fine pores, medium permeability.

Source: Directorate of Agriculture, GOWB.

Table 83: Physico-Chemical Properties of Soil Series "S3: Budhkarma, A1-A2-B2, Coarse Loamy" of Naxalbari Block, Darjeeling District

Horizon	*Depth (cm)*	*Mechanical Analysis*			*Textural Class*	*Chemical Analysis*			
		Sand (%)	*Silt (%)*	*Clay (%)*		*pH*	*CEC Me/100 gm*	*EC Ms/cm*	*Organic Carbon (%)*
Ap	0-22	40.6	38.8	20.6	L	6.3	7.9	0.10	1.17
A_1	22-38	46.6	36.8	16.6	L	6.6	6.3	0.05	1.09
A_2	38-111	68.6	20.8	10.6	SL	6.6	3.9	0.03	1.54
B_2	111+	62.6	26.8	10.6	SL	6.6	4.2	0.03	1.56

Source: Directorate of Agriculture, GOWB.

Table 84: S4: Soil Series (Nipnia, A-B, Sandy) Description of Naxalbari Block, Darjeeling District

Horizon	*Depth in cm*	*Description*
A_p	0-10	Diffused irregular lower boundary, dry soil colour brown (10YR5/3) with common medium faint mottles in reddish yellow colour (7.5YR6/8), sandy loam with medium sized stone and gravel (10 per cent), weak sub-angular blocky, fine to medium many pores, very fine to medium ample roots, medium permeability.
B	10-110+	Dry soil colour brown (10YR5/2), sandy loam with medium sized stone and gravel(10 per cent), weak sub-angular blocky, dry soft, many very fine to medium pores, very fine to fine ample roots, moderate permeability.

Source: Directorate of Agriculture, GOWB.

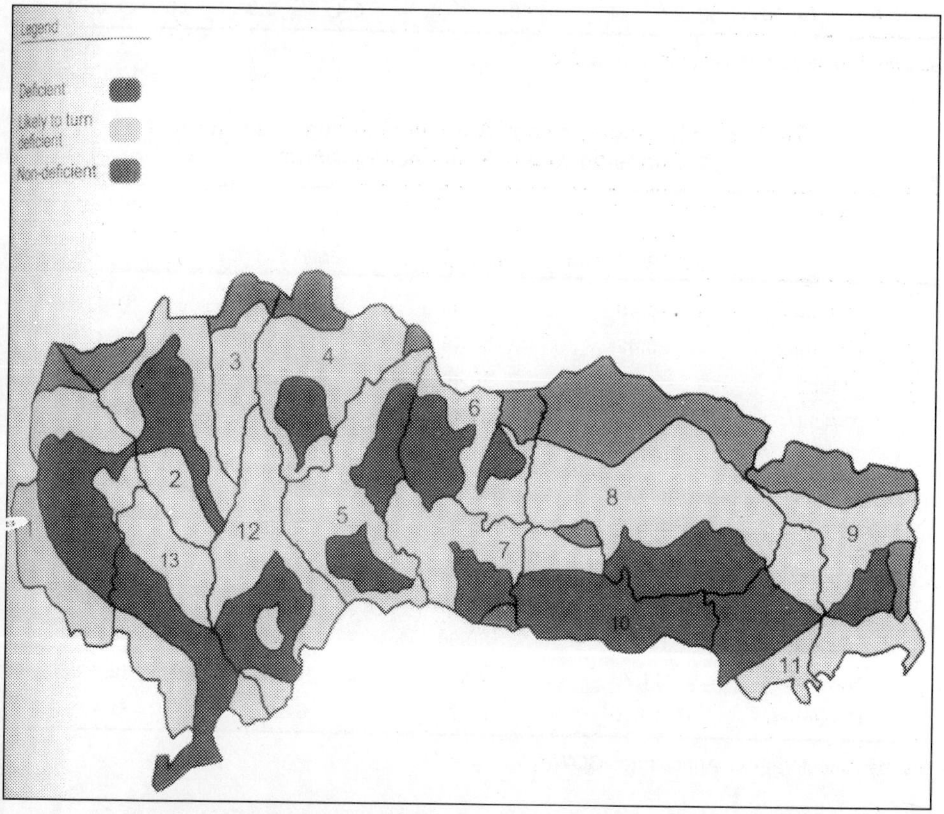

Source: Directorate of Agriculture, GOWB

Figure 17: Block-wise Status of Available Boron of Jalpaiguri District 1: Rajganj; 2: Mal; 3: Metali; 4: Nagrakata; 5: Dhupguri; 6: Madarihat; 7: Falakata; 8: Kalchini; 9: Kumargram; 10: Alipurduar II; 11: Alipurduar I; 12: Moinaguri; 13: Jalpaiguri

Table 85: Physico-Chemical Properties of Soil Series "S4:Nipnia, A-B, Sandy "of Naxalbari Block, Darjeeling District

Horizon	*Depth (cm)*	*Mechanical Analysis*			*Textural Class*	*Chemical Analysis*			
		Sand (%)	*Silt (%)*	*Clay (%)*		*pH*	*CEC Me/100 gm*	*EC Ms/cm*	*Organic Carbon (%)*
A_p	0-10	62.0	30.2	7.8	SL	6.6	7.3	0.14	0.62
B	10-110+	56.0	34.2	9.8	SL	6.4	5.9	0.06	0.74

Source: Directorate of Agriculture, GOWB.

Table 86: Physiography and Agroclimatic Regionalization of Phansidewa Block, Darjeeling District

Month	*Rainfall (05 Years Average from 1989-1994) (mm)*	*Temperature*		*Monthly Evaporation*
		Max. (C)	*Mini (C)*	
January	26.45	23.62	9.55	31.3
February	20.42	25.44	11.15	44.3
March	19.97	28.46	15.03	68.9
April	61.35	31.48	19.58	87.0
May	167.07	31.59	21.50	83.4
June	634.08	31.02	23.50	71.4
July	658.47	32.36	24.59	70.5
August	517.0	32.00	24.23	70.7
September	625.80	31.50	23.44	63.0
October	177.22	30.90	19.95	60.8
November	11.77	29.87	13.64	42.2
December	19.10	25.78	6.88	31.4

Source: Directorate of Agriculture, GOWB.

Table 87: Distribution of Land Under Different Soil Series Identified

Sl.No.	*Name of the Series*	*Abbreviation*	*Area in Ha.*	*Per cent Distribution to Total Surveyed Area*
1.	Leusipukuri	Lp	1381.38	35.37
2.	Radha	Rd	1683.25	43.10
3.	Bhuvita	Bv	840.81	21.53

Table 88: Soil Series Description (Leusipukuri) of Phansidewa Block, Darjeeling District

Horizon	*Depth in cm*	*Description*
Ap	0-25	Grayish brown (10YR5/2) and brown (10YR5/3), most sandy loam, massive, friable to slightly sticky, medium common pores, many fine roots clear wavy boundary. pH 4.3
$A1_1$	25-53	Brown(10YR5/3) to yellow brown (10YR5/4) with mottles of reddish brown (5YR4/4), loamy sand, massive, friable, slightly sticky, medium common pores with fine moderate roots, clearly smooth boundary,. pH 4.8
$A1_2$	53-73	Brown (10YR5/3), moist loamy sand, massive friable, slightly sticky, few to medium common pores, few fine roots, many medium mottles of yellowish brown (10YR5/8), clear wavy boundary,. pH 5.0
C1	73-132	In moist condition light grey (10YR7/2), loamy sand, very friable, non sticky, massive, fine to few fine roots, many medium detected mottles of yellowish brown (10YR5/8), clear wavy boundary. pH 5.1
C2	132+	In moist condition light grey (10YR7/1), coarse sandy, single grained, non sticky, uncommon pores, common medium faint mottles of yellowish brown (10YR5/8), no clear boundary, no appreciable special features. pH-5.1

Source: Directorate of Agriculture, GOWB.

Table 89: Physico-Chemical Properties of Soil Series "Leusipukuri" of Phansidwa Block, Darjeeling District

Horizon	*Depth (cm)*	*Mechanical Analysis*			*Textural Class*	*Chemical Analysis*		
		Sand (%)	*Silt (%)*	*Clay (%)*		*Organic Carbon*	*pH*	*E.C.*
Ap	0-25	71.4	21.4	7.2	Sandy loam	0.59	4.3	0.06
A11	25-53	67.4	25.4	7.2	–do–	0.26	4.8	0.03
A12	53-73	61.4	25.4	13.2	–do–	0.27	5.0	0.03
C1	73-132	81.4	11.4	7.2	–do–	0.10	5.1	0.04
C2	132+				Fine sand	0.09	5.1	0.04

Source: Directorate of Agriculture, GOWB.

Table 90: Soil Series (Radha) Description of Phansidewa Block, Darjeeling District

Horizon	*Depth in cm*	*Description*
A_p	0-24	Very dark grayish brown (10YR3/2), most sandy loam, massive to sub angular blocky, under moist condition friable and in wet condition slightly sticky, fine many medium common roots with presence of little pores, with clear wavy boundary, existence of few fine distinct yellowish red (5YR5/6) mottles.

Contd...

Table 90–Contd...

Horizon	Depth in cm	Description
A1$_1$	24-51	Brown (10YR5/3) under moist condition, silty loam soil, massive to angular blocky, friable, slightly sticky, presence of fine common roots, little or no pores, abrupt smooth boundary, presence of strong brown (7.5YR5/8), many medium distinct mottles. pH 5.7
II C1	51-84	White (10YR8/1) coarse sandy soil, single grained, loose under moist condition, slightly sticky, presence of few fine roots, no distinct mottles are present, soil have clear wavy boundary. pH 6.0.
IIIC1	84-129	Light gray fine sandy soil (10YR7/2) with many medium distinct mottles of yellowish red (5YR5/8), single grained structure, very friable under moist condition, non sticky consistency, no traces of roots, clear wavy boundary, pH 6.
IIIC2	129+	White coarse sandy (10YR8/1) soil, presence of common medium distinct mottles, of yellowish red (10YR5/8) colour, single grained structure, loose non sticky, no trace of roots and no clear boundary. Ph 6.3

Source: Directorate of Agriculture, GOWB.

Table 91: Physico-Chemical Properties of Soil Series "Radha" of Phansidewa Block, Darjeeling District

Horizon	Depth (cm)	Mechanical Analysis			Textural Class	Chemical Analysis		
		Sand (%)	Silt (%)	Clay (%)		Organic Carbon	pH	E.C.
A$_p$	0-24	55.4	31.4	31.2	Sandy loam	1.47	5.3	0.01
A1$_1$	24-51	66.0	34.8	9.2	–do–	0.34	5.7	0.01
II C1	51-84	69.7		12.8	–do–	0.03	6	0.01
IIIC1	84-129	77.4		11.5	–do–	0.04	6	0.02
IIIC2	129+				sandy	0.00	6.3	0.01

Source: Directorate of Agriculture, GOWB.

Table 92: Soil Series (Bhusivita) Description of Phansidewa Block, Darjeeling District

Horizon	Depth in cm	Description
Ap	0-31	Light brownish grey (10YR6/2) (dry) to grayish brown (10Y5/2) (Moist) with few fine distinct mottles of reddish yellow (5YR6/8), loamy sand, massive to single grained, very friable in moist condition, non sticky, presence of fine common pores, fine many medium common roots clear wavy boundary. pH 4.7
A1$_1$	31-55	Light grey (10YT7/3) (Dry) to grayish brown (10YR6/6) (Moist) with few fine faint brownish yellow (10YR6/8) mottles, fine sandy, single grained, loose, non sticky, non presence of pores with fine common roots, clear wavy boundary. pH 5.3

Contd...

Table 92–Contd...

Horizon	Depth in cm	Description
C_1	55-127	Brownish yellow (10YR6/6) (Moist), common medium distinct mottles of yellow colour (10YR5/8) fine sand, single grained, loose, non sticky, presence of fine common roots, with no pores, clear wavy boundary. pH 5.7
C_2	127+	White (10YR8/1) (moist) with common medium distinct yellowish red (5YR5/8) mottles, coarse sandy, single grained, very friable, non sticky, absence of clear boundary. This horizon has the special feature of presence of small to medium sized nodules, pebbles, iron and manganese concretious. pH 5.7

Table 93: Physico-chemical Parameters of Soil Series "Bhushivita" of Phansidewa Block, Darjeeling District

Horizon	Depth (cm)	Mechanical Analysis			Textural Class	Chemical Analysis		
		Sand (%)	Silt (%)	Clay (%)		Organic Carbon	pH	E.C.
Ap	0-31	78	14.8	7.2	Sandy loam	1.25	4.7	0.01
$A1_1$	31-55	100			Sandy	0.31	5.3	0.01
C_1	55-127	90	10		Sandy loam	0.09	5.7	0.01
C_2	127	100			Sandy	0.09	5.7	0.01

Source: Directorate of Agriculture, GOWB.

Table 94: Soil Fertility Status of Dakshin Dinajpur District

Name of the Block	O.C. %	Range of Nutrients Status and pH.			E.C.
		AVP 205	AVP 205	pH	
Balurghat Subdivision					
Kumarganj	0.05-0.70	12.225	40.600	5'4-7'5	0.01-1.4
Tapan	0.05-0.92	12-199	54-542	5'3-6'7	0.02-6.8
Balurghat	0.05-0.65	12-125	40-600	5'5-7'0	0.01-1.4
Hilli	0.05-0.08	12-125	40-500	5'5-7'0	0.01-1.4
Gangarampur Subdivision					
Kushmandi	0.02-0.58	4-98	36-206	4'0-6'2	0.05-0.26
Harirampur	0.17-0.39	8-52	8-78	4'9-6'3	0.02-0.12
Banshihari	—	—	—	—	—
Gangarampur	0.05-0.63	2-126	41-800	5'5-7'5	0.01-1.42

Source: Directorate of Agriculture, GOWB.

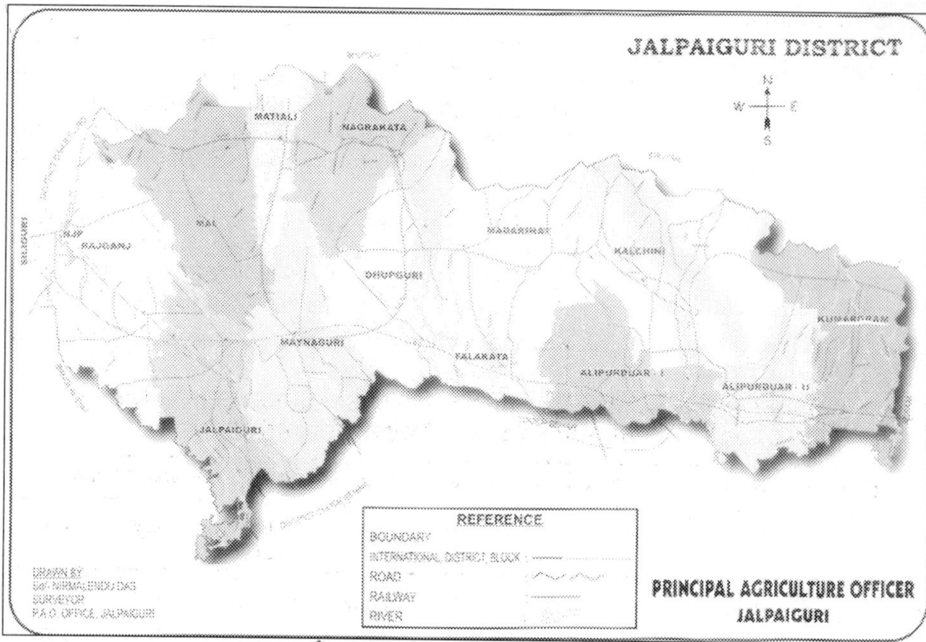

Figure 18: Block-wise Road, Railway and River Map of Jalpaiguri District

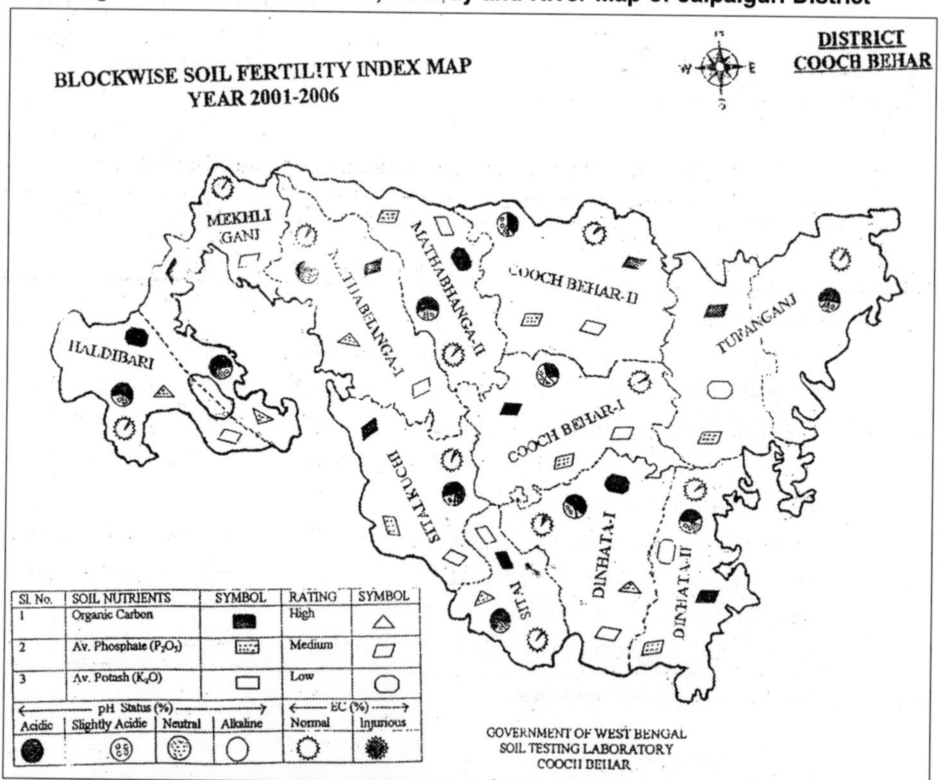

Source: Directorate of Agriculture, GOWB

Figure 19: Soil Fertility Index Map of Coochbihar District

Table 95: Soil Groups of Dakshin Dinajpur District

Name of the Block	Area Available for Cultivation	Soil Group			
		Predominant Group	% of Total Area	Other Groups	% of Total Area
Kushmandi	24630	Old Alluvial	70	Alluvial	30
Harirampur	29510	–do–	75	–do–	25
Gangarampur	25240	–do–	80	–do–	20
Kumarganj	23350	–do–	80	–do–	20
Tapan	35290	Laterite	75	–do–	25
Balurghat	500100	Old Alluvial	75	–do–	25
Hilli	7500	–do–	60	–do–	40
Total District	175620				

Source: Directorate of Agriculture, GOWB.

Table 96: Block-wise Land Situation of Dakshin Dinajpur District

Name of the Block	High Land	Med Land	Low Land	Total Land	Special Area Drought
Kushmandi	1000	11490	12140	24630	–
Harirampur and Banshihari	2050	15390	12070	29510	–
Gangaramnpur	2020	6970	16250	25240	1200
Kumarganj	1850	9620	11800	23350	–
Tapan	25690	5510	4090	35290	800
Balurghat	2050	25770	2280	30100	–
Hilli	905	5510	1085	7500	–
Total District	35565	80260	59765	175620	2000

Table 97: Block-wise Soil Types of Dakhin Dinajpur District

Name of the Block	Cultivable Sand	Area in Hect Sandy	With Differ. Loam	Soil Clay Loam	Clay
Kushmandi	—	540	2060	14190	7840
Harirampur and Banshihari	—	550	550	22590	5820
Gangaramnpur	1570	3040	3040	3040	14550
Kumarganj	1270	3870	3860	3860	10490
Tapan	21940	4950	1100	4300	3000
Balurghat	1610	4440	18250	5280	520
Hilli	380	4285	1060	1075	700
Total District	26770	21675	29920	54335	42920

Source: Directorate of Agriculture, GOWB.

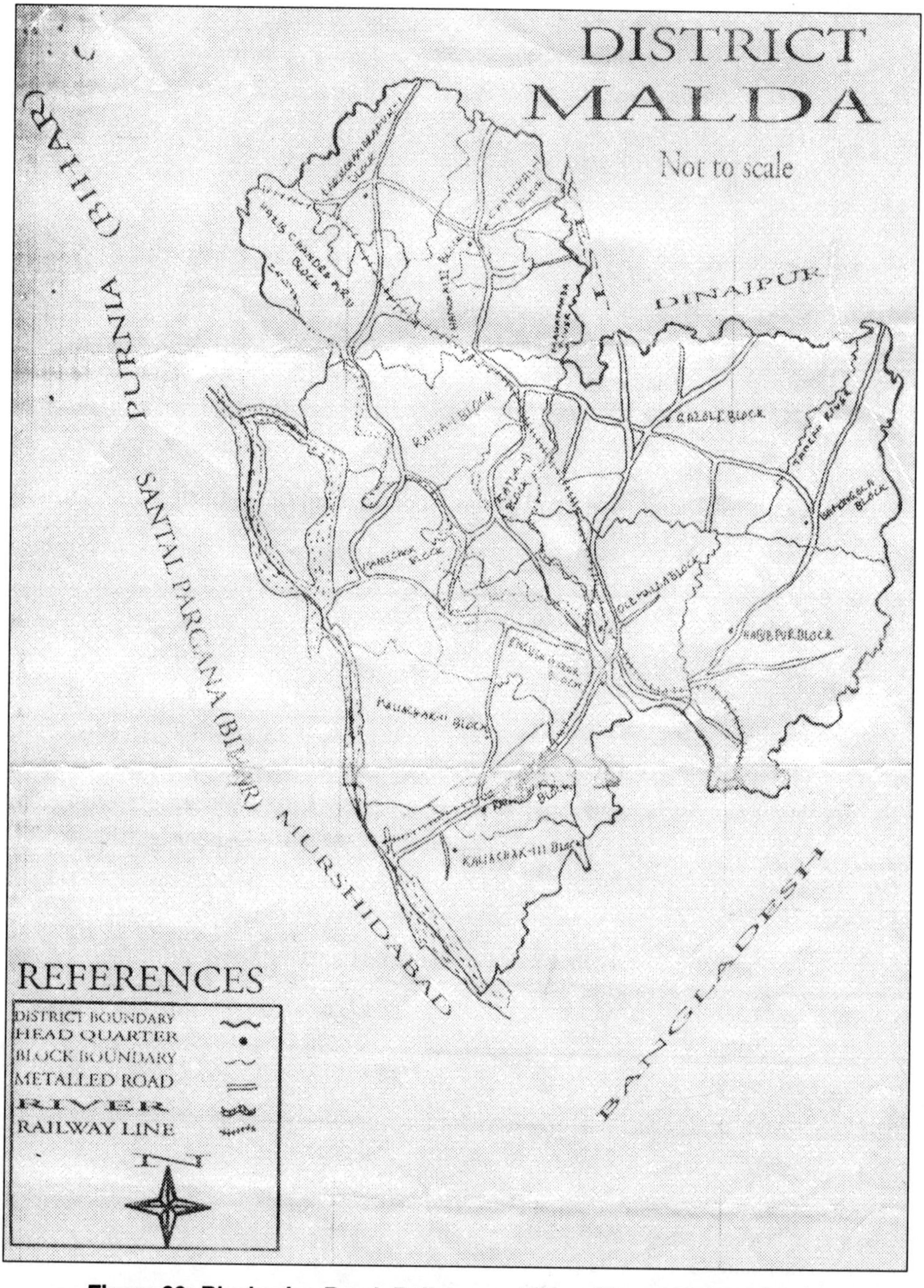

Figure 20: Block-wise Road, Railway and River Map of Malda District

Table 98: Soil Type (Texture) of Uttar Dinajpur

Sl No.	Name of the Block	Cultivable Area (ha.)					
		Sandy	Sandy loam	Loam	Clay loam	Clay	Total
1.	Chopra	13360	8100	400	400	–	22260
2.	Islampur	14005	10895	550	550	–	26000
3.	Goalpokhar-I	11650	10670	5880	650	650	29500
4.	Goalpokhar-II	11747	11323	5500	506	–	29076
5.	Karandighi	5710	11431	8560	3060	300	29061
Islampur Sub-division		56472	52419	20890	5166	950	135897
6.	Raigunj	–	31000	3200	600	400	35200
7.	Hemtabad	–	12860	3043	500	250	16653
8.	Kaliagunj	–	500	13060	9300	500	23360
9.	Itahar	–	1000	10882	17300	1000	30182
Raigunj Sub-Division		–	45360	30185	27700	2150	105395
Uttar Dinajpur District		56472	97779	51075	32866	3100	241292

Source: Directorate of Agriculture, GOWB.

Table 99: Soil Fertility Status of Uttar Dinajpur

Sl.No.	Name of the Block	O.C. %	Range of Nutrient Status			E.C. (m.mho/c.m.)
			AV. P_2O_5 (Kg/ha)	AV. K_2O (Kg/ha)	pH	
Islampur sub-division						
1.	Chopra	0.10–0.06	8.0–94.0	32.0–200.0	4.6–5.8	0.08–0.10
2.	Islampur	0.18–0.72	28.0–80.0	30.0–88.0	4.8–6.0	0.06–0.10
3.	Goalpokhar–I	0.18–0.02	28.0–85.0	36.0–210.0	5.0–6.0	0.04–0.10
4.	Goalpokhar–II	0.20–0.25	8.0–50.0	40.0–260.0	4.8–5.8	0.30–0.16
5.	Karandighi	0.20–0.30	9.0–60.0	40.0–290.0	5.0–6.2	0.08–0.20
Raigunj Sub-division						
6.	Raigunj	0.29–0.48	5.8–172.0	15.0–176.0	5.0–6.1	0.06–1.25
7.	Hemtabad	0.24–0.52	21.0–200.0	27.0–367.0	4.6–6.1	0.05–0.31
8.	Kaliagunj	0.18–0.38	8.0–73.0	63.0–81.0	5.0–5.6	0.04–0.07
9.	Itahar	0.18–0.90	4.5–89.0	12.0–75.0	5.2–6.3	0.05–0.29

Source: Directorate of Agriculture, GOWB.

Table 100: Soil Groups of Uttar Dinajpur

Sl. No.	Name of the Block	Area Available for Cultivation (ha)	Soil Groups			% of Total Area
			Predominent Group	% of Total Area	Other Group	
1.	Chopra	2260	New Alluvial	90	Alluvial	10
2.	Islampur	26000	–do–	90	–do–	10
3.	Goalpokhar-I	29500	–do–	90	–do–	10
4.	Goalpokhar-II	29076	Alluvial	60	New Alluvial	40
5.	Karandighi	29061	–do–	70	Old Alluvial	30
Islampur Sub-division		135897	–	–	–	–
6.	Raigunj	35200	Alluvial	80	Old Alluvial	20
7.	Hemtabad	16653	Old alluvial	65	Alluvial	35
8.	Kaliagunj	23360	–do–	60	–do–	40
9.	Itahar	30182	Alluvial	75	Old Alluvial	25
Raigunj Sub-division		105395	–	–	–	–
Uttar Dinajpur District		241292	–	–	–	–

Source: Directorate of Agriculture, GOWB.

Table 101: Block-wise Land Distribution of Uttar Dinajpur

Sl. No.	Name of the Block	High Land (ha.)	Medium Land (ha.)	Low Land (ha.)	Total (ha.)	Special Area (ha.)	
						Drought Prone	Flood Prone
1.	Chopra	14620	7296	344	22260	–	–
2.	Islampur	15010	9820	1170	26000	–	–
3.	Goalpokhar-I	12160	16800	540	29500	–	–
4.	Goalpokhar-II	13870	14426	780	29076	–	–
5.	Karandighi	6710	20995	1356	29061	–	–
Islampur Sub-division		62370	69337	4190	135897	–	–
6.	Raigunj	6450	27350	1400	35200	–	–
7.	Hemtabad	5200	10693	760	16653	–	–
8.	Kaliagunj	18400	18750	2770	23360	–	–
9.	Itahar	850	23597	5935	30182	–	1300
Raigunj Sub-Division		30900	80390	10665	105395	–	1300
Uttar Dinajpur District		76710	149727	14855	241292	–	1300

Source: Directorate of Agriculture, GOWB.

Table 102: Block-wise Micronutrient Data of Malda District

Sl.no.	*Name of the Block*	*Soli Quality of Pond*					*Remarks*
		pH	*Org. C*	N_2	PO_4	NH_4	
1.	Gazole Dev Bolck	6.0	0.52	2.01	2.2	4.0	
2.	Ratna-II Dev Block	6.2	0.60	1.95	2.3	4.2	
3.	Manikchak Dev Block	6.3	0.60	2.10	2.6	4.0	
4.	Old Malda Dev. Block	6.5	0.58	1.90	2.5	4.3	
5.	Kaliachak–II Dev Block	7.7	0.22	2.0	2.1	3.75	
6.	Kaliachak–III Dev Block	7.4	0.28	2.11	2.3	3.75	
7.	Habibpur Dev Block	6.5	0.28	1.75	2.1	3.65	
8.	Chanchal II Dev Block	7.5	0.15	1.68	2.3	3.80	
9.	English Bazar Dev. Block	5.5	0.20	–	30	–	
10.	Kaliachak I Dev Block	8.4	0.40	–	122	–	
11.	Harishchandrapur–I	–	–	–	–	–	TAL
12.	Harishchandrapur–II	–	–	–	–	–	TAL
13.	Chanchal Dev Block	–	–	–	–	–	
14.	Ratna I Dev Block	–	–	–	–	–	
15.	Bamongola Dev Block	–	–	–	–	–	

1. TAL–Harishchandrapur I& II, Chanchal I&II and Ratna I and II Dev. Block. Gangetic Alluvial Region.
2. DIARA–Manikchak, Kaliachak I, II, III and English bazaar Dev Block. Gangetic Alluvial Region
3. BARIND–Old Malda, Habibpur, Bamongola and Gazole. Dev Block Vindhya Alluvial Region.

Chapter 6
Decline of Indigenous Aqua Species

Ecology of the hill streams in general have considerably been deteriorated due to heavy land slides since 1968 resulting in the rise of river beds, disappearance of a good numbers of permanent pools harbouring game fishes. Despite of this rampage of nature they still hold enough prospects for development under planned programme and conservation. Fishes-young, adult and brood are found naturally trapped in stagnant transparent pools bounded by boulders (created by the river itself) often met their ruthless destruction by explosives at the hands of merry goers of picnic party/ poachers.

"Dealipara" is a process of wanton destruction of fish life in small streams by taking recourse to diversion of the flow of streams from its normal course thereby drying up the prospective pools having plenty of fish. Killing of fish with pesticides, copper sulphate, bleaching powder is also prevalent. Now a days the cheap availability of pesticides in the tea garden allures the delinquent to take recourse to such procedure for mass killing of fish. Often some quantity of copper sulphate tied in a piece of cloth is hung over a prospective pool to kill fishes.

Persons staying in remote rural areas use fruits of 'maidal' plant (*Erythrina arborescens*) or barks of "arare" creepers (*Acacia concinna*) to kill fishes in fast hill streams. All these types of wanton destruction of fish life are the main factors that can be attributed towards depletion of fish in hill streams.

The mahaseer in its fight for survival has to face much danger. In addition to the hazards faced by them through natural calamities such as landslides along with the river banks, floods washings faced a losing battle in its struggle for survival against its human foes, the poisoners and dynamiters.

With the availability of pesticides for agriculture and other purposes in the local markets and Government stores, some locals are having a field day in spreading poison in the still pools of the smaller off-streams killing off the fish. Fortunately, their activities are restricted as they can use poison only in the still pools of the off-streams and as such to that extent the mahaseer manage to escape this form of killings.

On the other hand, as a result of the tremendous activities in construction works near by, it become easy to procure explosives. Moreover, a large number of people had gained the expertise in handling the explosives having worked in the construction sites. The poachers therefore, lost no time and blasting with dynamites become more widespread in the region. The poachers indiscriminately blasted pools in the tributaries and off-streams flowing into the mainstreams of the Teesta. The indiscriminate use of dynamite kills all the fish in the pools. Countless numbers of fry and big fish are injured and killed out-right and as the biggest fish lie in the deepest part of the pools, they cannot be retrieved after blasting. This is wanton massacre of the mahaseer and the pools, after they have been blasted are literally emptied of all fish.

These dynamiters know the habit of the mahaseer and as such their activities are carried out in the pools of the tributaries and off-stream during the period of the upward and downward migration of the mahaseer during the month of February to May and thence from September to November when the pools teem with fish.

With the blasting of the pools in the tributaries and off-streams of the Teesta, the mahaseer is fighting a loosing battle for its survival. Nature cannot possibly make good the losses incurred by such massacres.

Uptil now the only way the mahaseer has been able to survive is because of the local superstitious belief of considering the mainstreams Teesta as the "Teesta Ma" the mother and sacred river on whose banks the dead bodies are cremated and their remains confined to its waters. The big pools in the mainstreams of the Teesta, the natural habitat of the mahaseer, have thus remained untouched and free from dynamiting. But for how long will tradition and belief prevail in these fast changing times?

Now a day, the people of hills have adopted an easy way of catching fish by using live wire with the help of portable generator. The method of electric fishing usually carried out during dry season *i.e.* during December to March. In this type of fishing, mostly smaller size of fishes, fingerlings of Katli, Mahaser, Asala, Garagoytola, Sulcatus are caught in a large quantity from very short interval of the river stream. This has become very popular among the river folks which in turn affecting the fish population of important species in this region.

The local people also mix poison (Thydene) in fish feed ball (mixture of rice bran, flour, ata, maize dust) and throw away into the rivers, pools, and wait in down stream for about half an hour to one hour time when the fishes in the said waterbodies starts dying during coming in the down stream. This is also one of the effective Wanton killing of fishes -both big and small, brooders and gravid.

Chapter 7

Scope for Preservation, Propagation and Conservation of Aquabiodiversity

The current chapter highlights the survey and restoration efforts of endangered fishes of West Bengal with special reference to North Bengal. A thorough survey was done on the aquatic resources in North Bengal in the District of Darjeeling, Jalpaiguri, North and South Dinajpur, Maldah and Coochbihar to identify the lesser known streams and rivers in those areas along with the major river system and wetlands. The study centers around the aquaculture sector in those districts, has been made on the endangered species of fish fauna of West Bengal, a total of 83 species have been identified as endangered. ARHMC unit of the Department of Fisheries has undertaken restoration programme of a number of fish species viz: *Ompak pabo, Notopterus chitala, Mystus tangra, Puntius sarana, Clarius batrachus, M. gulio, A. mola, A. testudeneous.* Successful breeding and propagation techniques of these fishes have been standardized by this center. This centre has also successfully experimented on hybridization of Indian Major carps (Rohu, Catla, Mrigala and Kalbasu) and the F1 progeny displayed the desired hybrid vigor characteristic. Finally this centre highlights the report on the current survey (*i.e.* during March 2005) and field trail of formulated feed in North Bengal. During this survey a number of interesting features came to our notice, which definitely needs a greater attention and research for biodiversity conservation in the North Bengal region.

Occupying 2.7 per cent geographic area of the country, the State of West Bengal possesses 10 per cent of the faunal and 20 to 45 per cent of the floral biodiversity. West Bengal Biodiversity Board (WBBDB) was constituted on September 16, 2004. Earlier, The Union Ministry of Environment and Forests passed the Biological

Diversity Act, 2004, with an objective to regulate access to genetic resources and associated knowledge by foreign individuals and Institutions and to ensure equitable sharing of benefits arising out of resources and knowledge available in the country. A three-tiered system of regulation is envisaged under the Act, which consists of the National Biodiversity Authority (NBA) at the head, followed by State Biodiversity Boards (SBB) and local level Biodiversity Management Committees (BMC). Every state of India, therefore, has a responsibility for the formation of the SBB that would be linked with the BMCs. No other Indian state is blessed with both coastal and coldwater aquatic resource within its borders, jhora fishery with roaring falls, flood plains, beels, baors, wetlands, sewage fed fishery, reservoir fishery, brackish water farming and capture fishery in river and sea. These diverse habitats have given rise to an array of aquatic species that call West Bengal home.

7.1. Background for Preservation and Conservation

Fishes are the most primitive vertebrates from which all other vertebrates including man has evolved. Fishes are cold blooded animals ranging in length from a few millimeters to twelve meters. They are adapted to an aquatic life with gills to breathe and fins to move about, although a few species can move on land and breathe air. They show a great diversity of form, habitat and are adapted to a range of environment. There are about 20,000 species of fishes which account for over forty percent of all living vertebrate species. Chemical fertilizers and pesticides are being used in agriculture. The wastes from Industry and agriculture find their place in the environment and pollute it, especially water. Most of the wastes ultimately go to the aquatic ecosystems. Apart from organic matter these wastes contain numerous harmful chemicals according to the report of United Nations Environmental Programme during 1974 four million chemicals were used. Since then thousand more chemicals are added in each year. The aquatic life is adversely affected due to the presence of these chemicals. Fish is the biological indicator of aquatic ecosystem health. If the fishes can survive and reproduce in a particular aquatic ecosystem, that ecosystem is considered to be a healthy one.

Soil erosion also affects the aquatic ecosystem. Due to soil erosion siltation takes place in river beds, which obstructs the flow of water. Due to this the spawning grounds of fishes and migration are affected which ultimately decreases the net production from these ecosystems. The lentic aquatic ecosystems (ponds, lakes, beels etc.) are also similarly affected by soil erosion.

7.2. Threats to Indian Biodiversity

India has a total of 89,451 animal species accounting for 7.315 of the faunal species in the world and the flora accounts for 10.78 per cent of the global total. The endemism of Indian biodiversity is high–about 33 per cent of the country's recorded flora are endemic to the country and are concentrated mainly in the North-East, Western Ghats, North-west Himalayas and the Andaman and Nicobar islands. However, this rich biodiversity of India is under severe threat owing to habitat destruction, degradation, fragmentation and over exploitation of resources. According to the Red list of Threatened Animals (IUCN, 2000), 44 plant species are critically endangered, 113 endangered and 87 vulnerable.

7.3. Conservation Practices in India and Indian Fish Biodiversity

In India conservation of wild life has been in practice since long back. In "Arthashastra" by Kautilya (300 BC) and the fifth pillar edict of Emperor Ashoka (242 B.C.) it has been mentioned that forest and wild life were protected by Royal Charter in ancient India. Since then there was a gap of about 2000 years. In 1935 the British Government convened a meeting to make a blue print for wild life conservation. Only in 1952 Indian Board of Wild Life conservation was formed.

'Hamilton (1822) and Day (1878) have contributed much on recording of the fish fauna of India. Some of the other important contribution towards the recording of fish fauna of India in this century are by Hora and Mukerji,1938; Jayaram, 1961, Misra, 1962 and Jhingran, 1985. Within India, Pakistan, Bangladesh, Burma, Sri Lanka, Nepal and Bhutan a total of 233 genera of with 742 species within 64 families of freshwater fishes have been recorded. The underlying causes of biodiversity loss, however, are poverty, macroeconomic policies, international trade factors, policy failures, poor environmental law/weak enforcement, unsustainable development projects.

Lack of perception about the criteria to be used for categorization of threatened fished coupled with a paucity of information has made fish conservation in India more difficult. Currently information on the categorization of different species of threatened fishes and the nature of threat at different geographical regions of the country have started to appear, for example Ponniah (1994) evaluated the existing system of designating threatened species categories and have proposed a modified set of criteria and has presented a simple categorization system along with a subjective way of assessing the various categories. Madhusoodane (1994) reported on the threatened fishes of river systems flowing through Kerala. Human activities have brought about major changes in riverine ecosystems of Kerala and there has been deterioration in the inland fishery resources. *Labeo dussumiari* which was once abundant in the river systems of Kerala now deserves the status and protection of an endangered species. The population of many species of fishes which once supported the fishery of Kerala has declined. The present practice of intensification of "Oothaeyijakkom", application of fish stupefying devices such as explosives, ichthyotoxic plants, poisons and pesticides and electrocution of fishes and other human imposed stresses such as over fishing, pollution hazards and engineering works have been reported to be some of the reasons of depletion of fishery in Kerala. Vats (1994) reported on the endangered fish species of Ganga basin, there is no comprehensive data available on the endangered fish species in Ganga basin, numerous authors, working on various aspects of fish and fisheries of different agro climatic zones in Ganga basin have reported different fish species which fall in the category of endangered species. The causes for decline of the fish fauna of Ganga basin has been attributed to various factors such as indiscriminate fishing, dynamiting, construction of dams and barrages, over utilization of water, forest denudation, mismanagement of farm lands, erosion of banks, point and non-point pollution and introduction of competitive exotic species. Roy (1994) reported on the fishes of West Bengal which been immediate conservation commercially important varieties of Elasmobranches, Clupeids, Siluroids, Gobids and Catfishes falls in this category. The status of Mahaseer in the different regions of the country has been given by various authors in the book entitled "Mahaseer the Game fish". There has been a general decline in the Mahaseer fishery throughout the country. Over exploitation, dynamiting, using phytotoxins, construction of road, soil erosion, deforestation and several other factors are thought to be responsible for this. Some existing measures for the conservation of fish population are, Control on overexploitation, Ban on dynamiting, Provision against pollution, Provision of mesh size, Close season, judicious introduction of exotic fishes.

An essential prerequisite to any programme of resource conservation is the proper taxonomic study of fish species occurring in the area concerned and a full check-list indicating the status of each species. Such a list would enable the IUCN to prepare an international list of endangered species to be included in the Red Data Book (Menon, 1996). But till date this list did not contain the endangered or threatened species of fishes of India.

The decline in the fish fauna can be attributed to several factors:

1. For commercial gains, only large sized fishes such as Indian Major Carps are preferred and the lesser native species are neglected.
2. Use of fish toxicants like mahua oil cake for eradicating so called weed fishes in commercial pisciculture.
3. Over exploitation and indiscriminate fishing of economically valued species.
4. Use of pesticides in adjacent agricultural land with subsequent introduction of these toxic substances into the water bodies through surface run-off.
5. Aquatic pollution resulting in alteration of water quality due to dumping of industrial and domestic wastes into the aquatic system.
6. Loss of breeding grounds due to siltation of large beel, baor and other fish habitat. Shrimp seed collection in coastal area is another factor.
7. Introduction of exotic species without proper research.
8. Dam construction and change of river course.
9. Conversion of wetland area into real estate business.
10. Plastic bags creating unaccountable damage to the habitat of aquatic resource

7.4. Efforts for Restoring the Current Situation

The steps should include for judicious exploitation of commercially important fish species, protection of the fish stock from pollution, restoration of the fish habitats and conservations of the rich fishery resources of the different districts:

1. Generation of self-employment and skill for interested farmers through demonstration and training.
2. Generation of public awareness regarding conservation of endangered fishes.

3. All-round habitat improvement by proper planning. Reclamation or re-excavation of the water bodies.
4. Prohibition on juvenile fishing by enforcing the use of proper gears and mesh size regulation.
5. Prohibition on fishing of brood fish by imposing closed season.
6. Through research and environmental impact analysis before introducing any exotic species.
7. Treatment of waste water before being let into the aquatic system.
8. Brood stock management and standardization of induced breeding and larval rearing techniques for the endangered species and restore the gene bank of threatened species by application of cryopreservation technique.
9. Tor species restoration, preservation and propagation – future plan of action from West Bengal Fisheries desk.
10. Technical support to the private hatchery owner for maximizing the quality seed production of endangered fish species.

Department of Fisheries, Government of West Bengal, have made a list of 83 species of vulnerable, threatened, rare, endangered and extinct fish. To making available of these fish species in their natural habitat, brood stock management, artificial propagation and larval rearing technique have been standardized of some endangered fish species like Chital (*Notopterus chitala*), Pholui (*Notopteres notopterus*), Pabda (*Ompok pabo*), tangra (*Mystus vittatus*), Nuna tangra (*Mystus gulio*), Saralpunti (*Puntius sarana*) etc. and extend the knowledge to the farmers up to grass root level.

7.5. Reasons for Selecting North Bengal Area as a Special Case

Three agro-climatic zones: Hill, Terai and Old Alluvial. We are to reach

1. 14.72 million people of which 86 per cent are rural gentry
2. Overwhelming number (over 80 per cent) of land owners belonging to smallest size class category (< 1 ha) followed by its immediately upper size class (1 -2 ha)
3. Nearly 40 per cent people under weaker section
4. Scores of ethnic groups like *Rajbanshi, Rava, Garo, Tota, Meche, Santhal, Madesiya, Lepcha* and other hill tribes.
5. The hungry thousands in growing number of closed and sick tea gardens.
6. Vast extent of flood prone, *diara and char,* and chronically problematic soil zones.
7. The entire region of North Bengal is resplendent with various hills streams and rapids which constituted a sizeable portion of the aquatic resources besides there are number of water bodies which happen to be the habitat of numerous local fish fauna which makeup a sizeable amount of ichthyofauna in the region of North Bengal.

8. 75 to 80 per cent ornamental fish export of India and best exporter of ornamental fish are from West Bengal state.
9. Now 70 per cent exportable ornamental fish coming from Terai region of West Bengal *i.e.* North Bengal.
10. About 70 per cent of the total freshwater fish variety of W.B. is available in N.B.

7.6. The Issues of Concern

- ☆ Declining contribution of primary sector (Composed of Agriculture, Forestry, Fishery, Mining and Quarrying) over Net SDP.
- ☆ Through the Fishery sub-sector appeared to be the sun shining one with a record growth (Table 103) of 6.83 per cent, the quick estimate is suggestive of the alarmingly low overall performance of the primary sector which was recorded to be only 1.23 per cent in 2004-05 and even negative growth was there (–0.98 per cent) for forestry sub-sector.

Table 103: Percentage Change of Growth in Different Sectors during 2003–04 and 2004–05

Sl.No.	*Components*	*2003–04*	*% Change*	*2004–05*	*% Change*
1.	Agriculture	20666.42	3.68	20779.044	0.54
2.	Forestry	555.95	4.52	550.50	–0.98
3.	Fishing	2647.73	3.96	2828.64	6.83
4.	Mining and Quarrying	24705.65	3.75	852.03	1.97
5.	Primary sector	24705.65	3.75	28010.21	1.23

7.7. The Problem of Soil in North Bengal

The soil of the North Bengal have also reached to an alarming state in many counts. Iron rich parent materials in some places caused very low soil pH while in other places acidity developed through leaching of bases due to high rain fall and partial oxidation of organic matter in soils. Due to this acidity problem, following pH related nutrient deficiency symptoms are often visualized.

- ☆ Low yield and nodulation in pulses
- ☆ Potassium deficiency symptoms in Vegetable and fruit crops.
- ☆ Boron Deficiency symptoms in many vegetable crops
- ☆ Phosphorus deficiencies in cereal crops
- ☆ The problem of acidity is magnified manifold in pond water of the entire region where fish growth and population commonly face drastic reduction.
- ☆ Nearly 55 per cent area is either deficient or likely to be deficient in Boron status
- ☆ 40 per cent area is either deficient or likely to be deficient in case of zinc.

- ☆ Whole region more or less suffer from deficiency in secondary nutrients like Ca and Mg
- ☆ Sulphur status is deficient or likely to be deficient in 39 per cent area.
- ☆ Comparable deficiencies in Macronutrient status exists all over North Bengal.
- ☆ Wide spread soil acidity appears to be organized from the parent material, climate and the organic matter status of the soils.

7.8. The Entire North Bengal still Remains in a State of Very Poor Rate of Farm Mechanization

There remains substantial presence of the weaker section and ethnic groups like *Raj Banshi, Rava, Toto, Meche, Garo, Santhals, Madeshia, Lepcha,* and other hill tribes, who constitute nearly 40 per cent of the total population of the North Bengal. Through livestock is the characteristic component of their livelihood system, this sector remains a grossly neglected one.

(*i*) Strengths

- ☆ Large extent of hill and terai forest belts
- ☆ Rich bio-diversity
- ☆ Large concentration of tea garden and tea industries
- ☆ Rich repository of medicinal and aromatic plants
- ☆ Wide nativity of fruits, flowers and orchids
- ☆ Large number of water bodies and river catchments
- ☆ Excellent average annual rainfall
- ☆ Varying climate to suit growing of different types of agri-horticulture crops
- ☆ Existence of two Agri-Export Zone (AEZ)
- ☆ Good communication network.

(*ii*) Weaknesses

- ☆ Subsistence state of agriculture
- ☆ Large proportion of weaker and vulnerable groups
- ☆ Poor production system efficiency leading to accrual of average daily wage rates even lower than the State average.
- ☆ Poor Human Development Index
- ☆ Poor harnessing of existing bio-resources including aquatic ones
- ☆ Less weathered soil deficient in a source of micro-nutrient
- ☆ Generally acidic soils having very poor water holding capacity
- ☆ Poor rate of farm mechanization
- ☆ Poor harness of rain water and micro-irrigation technology
- ☆ Extremely poor growth performance of livestock sector

- ☆ Weakly built R&D support infrastructure
- ☆ Extremely inadequate industrial growth
- ☆ Wide existence of flood prone areas

(*iii*) Opportunities

- ☆ Development of ancillary aquaculture based enterprises
- ☆ Large scale integration of crop-fish-livestock based farming system.

(*iv*) Vision

- ☆ Establishment of a network of e-kiosk based export advisory support system through capacity building and follow up technical support to young rural entrepreneurs

7.9. Critically Endangered Fish Species of North Bengal

The rational exploitation of animal genetic resources, including fish, is imperative for the improvement of socio-economic conditions of fishers of different districts of North Bengal. Legitimate concern about declining harvests and obvious reduction in bio-diversity of individual fish species, brewed over the course of time, and necessitated a more holistic approach to fisheries management and research related to both exploitation and conservation. India is fortunate to posses vast and varied fish germplasm resources. About 11 per cent (2,200) of the total world (more than 20,000) fin fish species have been recorded from the Indian subcontinent.

In the course of tracking peaks of development through rapid pace of industrialization and growth oriented projects, our beautiful aquatic endowments have recently been subjected to anthropogenic stress. Impact of such stresses can some times results in direct mortalities or cause alteration/destruction of habitats which in turn, interfere with life processes like feeding, migration and breeding etc. of the fish. Such alterations have been attacking the delicate balance of ties between the various strata of the ecological pyramid. Consequently, several fish species are exhibiting declining trends in different ecosystems and some have become endangered. Literature stands as witness to the exhaustive list of such adverse impacts and threats.

Comprehensive list of the threatened species of fishes is lacking. This lack of information on threatened species of fishes and the general lack of identification manuals of threatened species of fishes have combined to forestall the recognition and conservation of our vanishing fishes. An essential prerequisite to any broad programme of resource conservation is the proper taxonomic study of fish species occurring in the area concerned and a full check-list indicating the status of each species. Such a list would enable the IUCN to prepare an international list of endangered species to be included in the Red Data Book. Due to various anthropogenic stress like rapid industrialization, habitat destruction owing to river-valley projects, excessive water abstraction, siltation in rivers many due to a deforestation, over-exploitation, illegal killing of juveniles and brood fishes, injudicious application of pesticide, aquatic pollution, spread of dreaded disease and uncontrolled introduction of exotic fish, a number of fish fauna are now showing declining catch trend from the conventional fishing grounds and some have been even become threatened in West Bengal scenario, needing immediate conservation (Miller *et al.*, 1989: Master, 1990; Maitland, 1993; Dehadrai *et al.*, 1994; Kirchhofer and Hefi, 1996; Joshi and Raina, 1997; Shrestha, 1997; Ponwiah *et al.*, 1998). District-wise report on critically endangered fish species of North Bengal are depicted in Tables 104–108.

Table 104: Critically Endangered Fish Species of Uttar Dinajpur

Sl.No.	*Name of the Species*	
	Local Name	*Scientific Name*
1.	Punti (sarpunti)	*Puntius sarana*
2.	Pabda	*Ompok pabda*
3.	Kajali	*Ailia coilia*
4.	Kuchia	*Angulia bengalenis*
5.	Bagh arh	*Bagarius bagurius*
6.	Bacha	*Eutropiichthys vacha*
7.	Pabda	*Ompak bimaculatus*
8.	Koi	*Anabus testutdinius*
9.	Chital	*Notopterus chitala*
10.	*Pholui*	*Notopterus notopterus*
11.	Pangus	*Pangasias pangasias*
12.	Khoira	*Gudusia chapara*
13.	Raimas	*Raimus bola*
14.	Kursa	*Labeo gonius*
15.	Pankal	*Mastocembelus armetus*
16.	Tangra	*Mystus tangra*
17.	Tor	*Tor putitora*
18.	Tor	*Tor tor*
19.	Boal	*Wallgo attu*
20.	Nados	*Nandas nandas*
21.	Gourami	*Osphronemus nobilis*

Table 105: Critically Endangered Fish Species of Dakshin Dinajpur

Sl.No.	*Name of the Species*	
	Local Name	*Scientific Name*
1.	Nados	*Nandas nandas*
2.	Kursa	*Labeo gonius*
3.	Kuchia	*Amphibious kuchia*
4.	Tapa	*Tetradon cutcutia*
5.	Khalsa	*Trigogaster faciatus*
6.	Ghere machh	*Clupisoma gasua*

Table 106: Critically Endangered Fish Species of Darjeeling (DGHC)

Sl.No.	*Name of the Species*	
	Local Name	*Scientific Name*
1.	Tor	*Tor tor*
2.	Tor	*Tor puntitora*
3.	Scziothora	*Labeo dora*
4.	Asala	*Schizothorax richardsoni*
5.	Gooch	*Bagarius bagarius*
6.	Chepti	*Clupisoma Montan*
7.	Katli	*Neolissocheilus hexagonalepis*
8.	Gardi	*Labeo dero*
9.	Pahali sor	*Tor putitora*

Table 107: Critically Endangered Fish Species of Jalpaiguri

Sl.No.	*Name of the Species*	
	Local Name	*Scientific Name*
1.	Tor	*Tor tor*
2.	Tor	*Tor puntitora*
3.	Pholui	*Notopterus notopterus*

Table 108: Critically Endangered Fish Species of Coochbihar

Sl.No.	*Name of the Species*	
	Local Name	*Scientific Name*
1.	Koi	*Anabus testutdinius*
2.	Bagh arh	*Bagarius bagurius*
3.	Jewelled chameleon	*Badis badis*
4.	Bacha	*Eutropiichthys vacha*
5.	Gardi	*Labeo dero*
6.	Pabda	*Ompok pabda*
7.	Gourami	*Osphronemus nobilis*
8.	Khoira	*Gudusia chapara*
9.	Tor	*Tor tor*

West Bengal in bestowed by the gifts of nature, so far aquaculture is concerned. There is no dearth of wetlands, skilled farmers and qualified technologists in this state. Department of Fisheries, Government of West Bengal has been perusing to optimize the use of natural resources in order to elevate the socio-economic status of the fisher folk of the state. The secret of success in this work, lies on the successful percolation of technologies to the farmers–the most important ancillary of the project.

7.10. Role of Aquatic Resource Health Management Centre (ARHMC), Dept. of Fisheries, Govt of W.B. to Conserve these Species with the Following Objectives

Brood stock management, artificial breeding of rare species to make available of these endangered species in their natural habitat and restore the gene bank of threatened species by application of cryopreservation technique.

1. Identify the location of availability of endangered fish species in each and every District of West Bengal
2. To overcome the diseases problem during larval rearing tank and cultural ponds.
3. To generate income, self-employment and skill for interested farmers through demonstration and training.
4. Technical support to the private hatchery owner for maximizing the quality seed production.

7.11. Experiment on Standardization of Breeding and Larval Rearing Technique

Biological studies related to life-history traits are essential to understand the breeding strategies of the fish for standardization of the breeding and larval rearing techniques under captivity which are the viable aids in conservation and rehabilitation of the endangered species (Das and Pandey, 1999). Considering the maintenance of fish bio-diversity along with other biotic resources is a prerequisite for the well being of even human beings too (Smith and Chesser, 1981; Meffe, 1987; Pavlov, 1993) the strategies involved in artificial propagation for conservation of endangered fish species are briefly discussed.

The main objective of this investigation was to standardize the breeding and larval rearing technique of some endangered species of fish and extent the knowledge to the farmers up to grass root level as the fish keep available to the nature. With a view to making the conservation effort more effective, it has to be backed by an *ex situ* programme. The two main pillars of *ex situ* conservation programme are a) Live gene bank and b) Gamete/Embryo bank. In a live gene bank, the endangered species are reared in captivity, bred there in and genetically managed avoiding inbreeding depression, domestication and unintended selection. In gamate/embryo bank, adequate samples representative of the natural genetic variation of endangered species are held in an suspended state of animation under liquid nitrogen. In the first phase of "Endangered fish species breeding" programme, we have selected some freshwater fish, reared them in captivity for maturation, artificially propagate them for larvae production and finally rear the larvae in controlled condition. Experimental result of some fish species are given in Table 109.

Table 109: Experiment on Some Fish which are Going to Disappear from their Natural Habitat

Fish Species	*Hormone Used*	*Sex Ratio Male : Female*	*Dose of Hormone Used (per Kg. Body Weight)*	*Percentage of Hatching*
Pabda, *Ompok pabo*	Pituitary extract, Ovatide	2:1 2:1	16 mg (in female only) 3.0 ml (in female only)	50 per cent 70 per cent
Chital, *Notopterus chitala*	Pituitary extract	3:1	11 mg (in female only)	60 per cent
Pholui, *Notopterus notopterus*	Pituitary extract	2:1	4 mg (in female) 2.5 mg (in male)	70 per cent
Indegenous tangra *Mystus vittatus*	Ovatide	2:1	8 ml (in female)	80 per cent
Saral punti *Puntius sarana*	Pituitary extract	2:1	4 mg (in female) 2.5 (in male)	70 per cent
Indegenous magur *Clarius batrachus*	Ovaprim	1:1	2.5 ml in female 1.0 ml in male	50 per cent
Nuna tangra *Mystus gulio*	Ovaprim	2:1	2.5 ml in female 1.0 ml in male	80 per cent
Mourala *Amblypharingodon mola*	Pituitary extract	2:1	4 mg (in female) 2.5 (in male)	80 per cent

The breeding technique of two endangered fish 1) a freshwater endangered fish Pabda, *Ompok pabo* and 2) a brackish-water endangered fish Tangra, *Mystus gulio* are discussed here.

Pabda, *Ompok pabo* (Hamilton) and Tangra, *Mystus gulio* constitutes important component of riverine and brackish-water fisheries in the Indian sub-continent. Sporadic informations are available on the biology and culture practices of these two species. *Ompok pabo* and *Mystus gulio* are the Indian major catfishes belonging to the family siluridae. Pabda dwells and breeds in the rivers and reservoirs and in connected water sheds in flooded condition and Tangra found in seas and estuaries throughout the India. *Mystus gulio* dwells and breeds in the estuaries during monsoon. Report of captive breeding on these two species is not available so far from the Indian sub-continent and abroad till date.

(*i*) Breeding Technique and Management of Pabda, *Ompok pabo*

Pabda, *Ompok pabo* Juveniles were collected from the river Punarvava, in the vicinity of Malda District, West Bengal. They were stocked in a polyculture tank with Indian Major Carp in a 0.5 h. pond. Along with regular liming the fish stock were fed with conventional feed consisting of mustered oil cake and rice bran in the ratio of 1:1. Pabda was maintained throughout the carp culture period. The brooders attained maturity after one year and the average body weight were recorded to be 85g. In the month of May both female and male brood fish were found to be gravid.

Adults showed sexual dimorphism. Matured males exhibited roughness of the first ray of pectoral fin at the lower side and possessed a narrow and rather pointed genital papilla, along with free oozing white milt on slight pressure on the abdomen. In case of female the pectoral fin was smooth and the genital papilla was found with thick muscular round the opening.

The spawners were selected for induced breeding experiment in the month of July, 2000. The maturity status of female was assessed on simple cathet ring device. Free oozing males and ripe females were used in the ratio of 2:1 respectively for breeding.

Breeding were done in a bundh. Bundhs are special type of impoundments where riverine conditions are simulated during monsoon months. The bundhs after a heavy shower, receive large quantities of rain water from their extensive catchment and provided large spawning ground (Jingran, 1982). Brooders were collected from brooders tank and kept in a hapa in the bundh and conditioning for 24 hrs. Average weight of brooders (80 ± 5g) were selected for experiment.

For the breeding operation, two experimental set up, one with carp pituitary extract and second with a synthetic hormone 'ovatide' (Hemmo pharma, Mumbai, India) were used. For each experiment six male and three female were kept in a separate breeding hapa. Single doses of hormone were administered. No male fish were injected. The dose of hormone was used, as pituitary 16mg/kg body wt. and Ovatide was 3.0 ml/kg body weight. Before using, the synthetic hormone was diluted by double distilled water. The main water quality parameters of the brooders tank, breeding tank and larval rearing tank were recorded regularly (Table 110) following the method given by APHA (1995).

Table 110: Major Water Quality Parameters in the Brooders, Breeding and Larval Rearing Tank for Pabda, *Ompok pabo*

Pond	*Temp (°C)*	*pH*	*DO (ppm)*	*NH_4N (ppm)*	*Alkalinity (ppm)*	*Hardness (ppm)*
Brooders Tank	29±1	7.5–8.0	6–6.5	–	130±10	110±15
Breeding Tank	30±1	7.6–7.9	7–7.6	–	140±15	112±10
Larval Rearing Tank	29.5±1	7.5–7.8	6.9–7.2	–	135±15	115±10

Fertilized eggs were found free and demarsal. Water hardened eggs resembles small and measured (1-1.5 mm.) in diameter. The number of eggs was estimated by volume and then transferred to the hatching hapa. Females were examined to ascertain wheather spawning was complete or partial. The completely spawned fished were released in the brood fish pond after giving a dip in a metheline blue solution to avoid any possible infection.

After 22 hours of fertilization, embryo hatched out. Newly hatched embryo were 5-6 mm in length, small yolk sac attached with the hatchlings. After 24 hours of hatching, larvae were then transferred from the hatching hapa to the nursery tank for further development and agent kept in another hapa. Larvae were fed with formulated feed after three days of hatching. Formulated feed were prepared by steam cooking of the mixture of egg and mussel (Karmakar *et al.*, 2000). "Piscimix" powder was mixed with the prepared feed at the rate of 10mg/kg feed. The mixture then sieved according to necessary mess size and thoroughly washed in tap water.

Among the Synthetic hormone (ovatide) and carp pituitary extract ovatide gave the best result (Table 111). The number of egg release with the ovatide induced hormone was (70 per cent) higher than that of pituitary extract (50 per cent). Hatching percentage was higher in ovatide and 28 per cent higher than the pituitary extract.

Table 111: Larvae Production of Pabda, *Ompok pabo*

Set	*Hormone Used*	*No. of Fish*	*Dose of Hormone (per kg. body wt.)*	*No. of Egg Released*	*Percentage of Hatching*
I	Pituitary gland extract	Female = 3 Male = 6	16 mg. –	4200	50 per cent
II	Ovatide	Female = 3 Male = 6	3 ml. –	20500	70 per cent

(*ii*) Breeding Technique and Management of Tangra, *Mystus gulio*

Adult tangra, *Mystus gulio* were collected from the local bheri at Digha, Purba Midnapore, West Bengal and were stocked in a round cemented tank in hatchery for the period of 15 days for acclimatisation. Along with regular water exchange the fish stock were fed with conventional feed consisting of mustered oil cake, rice bran, dry fish in the ratio of 1:1:5. Adults showed sexual dimorphism. Males exhibited a muscular, conical reddish pink genital papilla just above the genital opening along with free oozing white milt on slight pressure on the abdomen. In case of female the genital papilla was found with thick muscular round the opening.

The brooders were selected for induced breeding experiment in the month of June, 2001. Total experiment was carried out at "Digha Experimental Prawn Hatchery", Digha, West Bengal. The maturity status of female was assessed on simple cathet ring device. Free oozing males and ripe females were used in the ratio of 2:1 respectively for breeding.

Breeding were done in 200 liter round earthen cistern (2 feet diameter). Total 9 number of earthen cistern were taken and filled with saline water (20 ± 1.5 ppt) collected from adjacent creek. Suitable water quality parameters were maintained through out the period of experiment. Vigorous aeration was provided throughout the experiment period with blower. Brooders were collected from brooders tank and kept in cistern at the ratio male: female = 2: 1. Two male and one female were introduced in each cistern. Average weight of brooders (female, 110 ± 10g; 15±1.5 cm. and male 160 ± 15g; 20±2 cm.) were selected for experiment.

For the breeding operation, a synthetic hormone 'ovaprim' was used. Double doses of hormone were administered to the female. Male fish were injected for single dose at the time of final dose to the females. The dose of hormone were used, 2.5 ml/kg body weight in case of female and 1.0 ml/kg body weight in case of male. Before using, the synthetic hormone was diluted by double distilled water. The main water quality parameters of the brooders tank, breeding tank and larval rearing tank were recorded regularly (Table 112) following the method given by APHA (1995).

Table 112: Major Water Quality Parameters in the Brooders, Breeding and Larval Rearing Tank for Tangra, *Mystus gulio*

Pond	*Temp (°C)*	*pH*	*DO (ppm)*	*NH_4N (ppm)*	*Alkalinity (ppm)*	*Hardness (ppm)*
Brooders Tank	29 ± 1	8.0 – 8.2	7– 7.5	-	130 ± 10	20 ± 3
Breeding Tank	29 ± 1	7.5 – 7.9	6.8 – 7.0	-	140 ± 15	20 ± 2
Larval Rearing Tank	30 ± 1	7.9 – 8.0	6.5 – 7.0	-	135 ± 15	20 ± 2

After 10 hours of injection first batch egg (about 20 per cent) were released, second batch (about 35 per cent) were released after 2 hours. Rest amount of eggs (45 per cent) were released after 3 hours. Fertilized eggs were found free and demarsal. Water hardened eggs resembles small and measured (2.2 – 2.4mm.) in diameter. The numbers of eggs were estimated by volume and then transfer to the hatching hapa. Females were examined to ascertain wheather spawning was complete or partial. The completely spawned fished were released in the brood fish tank after giving a dip in a metheline blue solution to avoid any possible infection.

After 20 hours of fertilization, embryo was hatched out. Newly hatched embryo was 3.5 – 4.2 mm in length, small yolk sac attached with the hatchlings. After 24 hours of hatching, about 80 per cent of yolk sac was absorbed and after another 12 hours hatchlings were free from yolk sac. Larvae were then transferred from the hatching hapa to the nursery tank for further development. Larvae were fed with *Artemia* nauplii after complete absorption of yolk sac.

Table 113: Larvae Production of Tangra, *Mystus gulio*

Set	*Hormone Used*	*No. of Fish*	*Dose of Hormone (per kg. body wt.)*	*No. of Egg Released*	*Percentage of Hatching*
I	Ovaprim	Female = 9 Male =18	2.5 ml 1.0 ml	30000	80 per cent

It is apparent from the above experimental result with Pabda, *Ompok pabo* and Tangra *Mystus gulio* that the experiment was successful with the synthetic hormone, encouraging result were found by carp pituitary extract in case of pabda.

Feeding is the most important factor in the larval period rearing. In Pabda, formulated feed with the mixture of piscimix helps to combat mortality, enhances survival rates, allows larger growth. It also checks mal-nutrition, Piscimix mixture helps to maintains steady growth of bones and takes care to body muscles.

However, the induced breeding of *Ompok pabo* and Tangra *Mystus gulio* are comparable to the induced breeding of carp but need special attention for larval rearing, mostly after 24 hours of hatching there are large mortality in the larval population. Some toxicity appears due to deterioration of water quality parameter in the hatching hapa by decomposition of egg shells. But after the mortality the rest population of survived larva attained fingerling without facing any problem.

Attempts to breed the endangered fish species artificially help not only to maintain the aquatic bio-diversity but also help to make proper utilization of vast aquatic resources. In recent future, are going to breed some other endangered fish species like Nados, *Nandus nandus;* Parse, *Mugil persia;* Bhagan, *Mugil tade;* Bhetki, *Lates calcerifer,* Pangus, *Pangasius pangasius* etc. Breeding technology of these species in captivity is not available. An improved breeding technique by using bio-filter in hatchery, making suitable feed formulation for larval rearing expected to help to increase the survivability of fish larvae. But this entire attempt will get zero result without public awareness. Hence it is the high time for creating awareness among the aquaculturist, agriculturists, scientists, professionals, and everybody associated with fisheries.

To make successful the objective of endangered fish species in West Bengal some potential spot in few Districts are idenified where these fish species are available and are idenified for easy identification of locality where we can implement our future plan in this regard will be mapped.

With a view to making the conservation effort more effective, gamate/embryo bank, adequate samples representative of the natural genetic variation of endangered species are held in a suspended state of animation under liquid nitrogen.

While a gamete/embryo bank in comparison to live gene bank is a better cost effective strategy, the present state of technology allows only cryopreservation of fish spermatozoa. Successful cryopreservation of fish eggs and embryo are yet to be achieved.

7.12. Scope for Preservation and Conservation of Mahser

In order, to protect the mahaseer, topmost priority must be given to put a stop to the massacre of the fish by dynamiting. Poaching with hooks and nets is tolerable. The State Forest Department and Fish and Wild life Department of the Government Sikkim and State Fisheries Department have to take immediate measures towards this. Whenever the mainstream of the Teesta or its tributaries and off-stream force the boundary between the two States, the respective banks of the mainstreams or tributaries and off stream should be patrolled by the Government agencies having the administrative jurisdiction of the area. Such patrolling should be undertaken as an intensive scale during the period from February till the end of May, thence again during the month of September, October and November.

The poacher caught dynamiting should be arrested and given the maximum penalty which would set an example and act as a deterrent to other dynamiters. In addition to this measure, it is suggested that genuine anglers be asked to keep a sharp look up on such dynamiters and their activities reported immediately to the nearest Government outposts. It is further suggested that villagers living near the tributaries and off-streams be given incentives to detect the dynamiters by offering them moneytory or other rewards leading to the arrest and punishment of the poachers.

Dynamiting of fish should be taken as a serious offence as define under the explosives act of the Government and the punishment assigned to those crimes should be fully imposed to the offenders. Moreover, the use of explosives is also dangerous to the users themselves. The Government should, therefore, take account of this state of affairs. Wide publicity should be given against the use of explosives and public opinion should be raised against the indiscriminate use of explosives.

If the dynamiters of the tributaries and off-streams can be put a step, then nature herself could be able to replenish the stock of Mahaseer. The Mahaseer would again team in countless numbers in the Teesta and her tributaries making the Teesta basin truly the paradise for anglers, when the mahaseer would reign supreme once more, as king of our mountain rivers bringing thrills and excitement galores to the sporting fraternity of anglers.

In the Scheme "Operation Mahaseer" it is, therefore, suggested to protect the second phase of the river Teesta *i.e.* the stretch from Rangpoo to Sevoke by posting of Conservation Assistants at the spots found out of survey.

In these spots, Department of Tourism may be requested to construct "FISH LODGE" for the accommodation of anglers and tourists. Mention be made here that the Government of Sikkim has already launched a similar scheme on their side to bring to book the law infringing *and* hostile elements engaged in the menace of explosive, poisoning and wanton poaching in the rivers. The Co-ordination of Forest Department and Police should also be sought for prompt registration of cases.

7.13. Himalayan Mahseer Fingerling Production for Conservation

It is now widely accepted that mahseer has dwindled in almost all types of habitats. It has rightly been declared as endangered by the experts in several seminars/ symposia/conferences and workshops. There can be only two ways for increasing its numbers. One is to allow the fish to propagate itself helped Mother Nature. The second alternative is to go in for large scale seed production of mahseer and utilize the seed for repopulating its habitats. Here, it must be made clear that agroclimatic conditions and mahseer species in Himalayan belt are entirely different from those in Western Ghats. Water temperature in Himalayan belt goes down below 10°C. When it happens the water becomes lethal for golden mahseer. In contrast, Deccan Mahseer never faces such extreme conditions. In the Himalayan belt this happens often. Because of this problem, the policy for the production of golden mahseer has to be two pronged. One should take care of its seed production and the other for its production for consumption. There are a large number of natural as well as man made water bodies in upland States. They deserve to have mahseer hatcheries where its seed can be produced in good numbers and stocked in the water bodies of the area. The National Research Center on Coldwater Fisheries, Bhimtal has developed and refined the design of a flow–through hatchery which can be set up for producing the seed and later dismantled at a very short notice when needed. A hatchery of this kind can produce about a quarter million of seed at a time. NRCCWF is helping user agencies to establish these units by providing technical assistance. Serious efforts and attempts are required to save "Mighty Mahseer" in Himalayan region. A holistic approach on the lines of "Save Tiger Project" is required for which an implementable policy is urgently needed. We must move ahead with greater responsibility and dedication in this direction.

7.14. Remedial Measure Taken by the Department of Fisheries, GOWB

As per the instruction of Hon'ble MIC Fisheries Department Govt. of West Bengal a decision was taken in a meeting held in the Chamber of the Director of Fisheries, GOWB on 14.03.2005 that hence forth only endangered fish species seeds would be supplied in the social fishery programme in different districts. (Director of Fisheries memo no.-FTR 73/14 dt. 16.03.2005), Also decision was taken for establishing hatchery in every district of West Bengal for breeding of endangered variety of fishes and native variety of magur (*Clarius* sp.)

Solution

1. Maintaining of breeding grounds and rearing grounds both in rivers and beel baors.
2. Effecting dredging of water beds of all rivers.
3. Restoration of the ecosystem of beel, boors.
4. Stopping exhaustive netting of water bodies.
5. Base of dumping of raw sewage. Only treated sewage of stipulated quantity to be released in nature.
6. Creation of sanctuaries for aqua species. Like national parks of forestry for breeding and rearing of endangered species and relocating then in this original habitat in nature.

7.15. Measures to be Adopted for Conservations of Aqua Species

Aquaculture and aquatic resource are basically the victims of pollution. Changes in the biological, chemical, geological or physical component of ecosystems can have impacts on resource population and community. Effluents from factories, sewage from municipal drainage, fertilizers and pesticides from agricultural wash offs happen to defile our aqua farms and equate resources. There are large scale fish kills and disease due to water pollution in one form or the other. Under such circumstances the fist crop insurance policy should have a clause regarding "polluters pay" whereby the pollution causing units can be brought to book and fined accordingly As most of our rivers have virtually been turned into glorified drams it is high time a common central guide line be drawn for effluent discharge which will specify the physicochemical and bacteriological standards of safe effluent which can be discharged in open water. The main causes for extinction of different species in this area may be as follows:

1. Dams and barrages are constructed on the river Teesta that hampers the migrations of migratory fishes of Teesta especially the Mahaseer.

2. Human settlements along the river Teesta and its tributaries are causing pollutions which are effecting the movement of fishes during the breeding time.
3. The water quality analysis reveals that water temperature ranging from 10-20°C, air temperature from 13-25°C, pH 7.6-8.2, D.O.–10.5 -13.6 mg/l, Specific conductivity 56-85 mohs. The water is slightly turbid due to the high content of sand particles during rainy seasons.
4. New species of loaches have been found namely-*Nemacheilus botia, N. devidevi, N. guntea* which has great ornamental value. Others like gourami, Vitti (*Danio* sp,). Some snake heads (*Channos* sp.) were also found.
5. Some illegal methods of catching fishes like dynamiting, poisoning, hammer fishing, electro fishing etc are also observed, which is also causing a decline in the coldwater fish population.
6. Fish Farm should be set up for the breeding of indigenous fish species in captivity, the produce to be released in the natural habitat (Ranching).
7. Some of the government officials should be provided training exclusively in the field of breeding, nutrition, hatchery management and diseases of coldwater fishes to man the seed farms.
8. Training programme to be imparted to the beneficiaries on coldwater fish culture.
9. Suitable scheme may be formulated for the coldwater fish farmers and provide them with suitable financial assistance from various institutions.
10. Awareness camps should be organized on a regular basis through department, N.G.O's, for which a substantial amount should be earmarked in the budget.
11. Angling should be allowed with permit; for which an authority to be set up.
12. There should be a task force for strict vigilance to stop illegal fishing.
13. The cultural technologies for the coldwater fisheries should be standardized and made available to the fish farmers.
14. A coordinated approach towards this need to formulate with the Fisheries Dept, Govt. of Sikkim.

And towards this the National Research Center on Coldwater Fisheries (ICAR) may take a joint venture with State Fisheries dept. and they may set up one sub center in the region. To rejuvenate endangered fish species immediate steps to be taken for mass production of seed and stocking of juveniles in their natural habitat of these rivers and streams. For this purpose there is a need to set up coldwater fish farm in the hill area. It will play a vital role for propagating conservation and culturing endangered species and side by side will also help in producing a healthy commercially important disease free important species that can be raised in their natural habitat. This will also generate new hope for the near extinct important sport fishes. The fisherman of the locality who is dependent on the fish catch will also be benefited. There is also tremendous scope for augmenting the fish production from the jhora fish pond by manipulating fish combinations of Indigenous species like mahseer (*Tor putitora*), snow trouts (*Scizothorax sp*), Katli (*N. hexagonalipes*), Common carp (*Cyprinus carpio*), Grass carp (Ctenopharyngodon *Idella*). In few pockets of Darjeeling areas the farmers have already started using the above combinations, not in scientific manner yet results shows increase in fish production. Hence, if scientific fish culture is adopted for the coldwater fishes, the jhora fishery scheme will be lucrative venture for the local residents.

Govt. of West Bengal with the help of D.G.H.C. has set up fish seed farm at Reyang in Kalimpong sub division to cater the demand of fish seeds mainly grass carp and common carp to the fish farmers of the hills and to take up preliminary study of the coldwater fish species on their captive breeding, food and feeding habit etc. There is an urgent need to start up with the scientific approaches toward the assessment of present fishery status of rivers and jhoras. Importance is also to be given to the thorough study of biodiversity, water quality, physico-chemical parameters of different streams to come up with suitable recommendations for the future ventures. Fisheries Department, Govt. of W.B. has started preliminary work towards this object but this only can take a shape towards transcription into reality with an ultimate aim, if the following concerned line department/authority will act together with a common objective:

- ☆ W.B. State Forest Deptt.
- ☆ W.B. State Tourism Deptt.
- ☆ Sikkim Forest and Fisheries Deptt.
- ☆ National Research Center on Coldwater Fisheries (ICAR);
- ☆ Central Inland Fisheries Research Institute (ICAR);
- ☆ National Fish Bureau Genetics Resources (ICAR);
- ☆ Zoology Dept, North Bengal University; and
- ☆ Faculty of Fishery Science, West Bengal University of Animal and Fishery

If all should join their hands together with the West Bengal State Fisheries Dept.'s initiative for a concerted effort. And towards this the first step would possibly be to form an Advisory Committee under the chairmanship of the Fisheries Secretary, Govt. of W.B. and the National Research Center on Coldwater Fisheries (NRCCWF) to set up one sub center in this region in collaboration with the State Fisheries Dept.

Study of the social, economic and institutional aspects of fisheries of North Bengal with special emphasis to coldwater area of the state of West Bengal is a recent need. Despite a sound general understanding, detailed knowledge of the biology and ecology of individual species and ecosystems remains poor. Further studies on individual species, communities and ecosystems are urgently needed. Understanding of the social organization of the fishery and the relationship between fisheries and other livelihood strategies should be given prior importance.

Chapter 8

Ornamental Fish Species for Commercial Purpose

India is the second most populous country in the world, with a 2003 total population of 1.068 billion, representing 17 per cent of the world's population, occupying 2.4 per cent of the world's land mass. The state of West Bengal is the fourth most populous state situated in the Eastern Region in India. 82 million people stays in the State. It is 2.7 per cent of India's area but about 7.8 per cent of the country's population, this state ranks first in terms of density of 904 per sq. km as per the 2001 census. The boundaries of the state are Nepal, Bhutan and the Sikkim on the North, Goalpara district of Assam and Bangladesh on the East, Orissa and Bay of Bengal on the South and Bihar on the West. The proportion of the rural population in the state stands at 72 per cent. According to the Planning Commission, the proportion of population below the poverty line in 1999-2000 in West Bengal was 31.85 per cent. The percentage of Schedule Caste (S.C.) and Schedule Tribe (S.T.) population are 28.6 and 5.8 respectively in the rural areas. It may be noted that S.C., S.T. and Minorities together account for more than half the population and they are the poorest groups in rural Bengal. A significant part of the state is relatively more economically backward and also tends to be less advanced in terms of human development. Gender discrimination is common throughout the region. Women in the area commonly experience a multitude of disadvantages and poorly represented in policy and decision-making. For women, the problems are compounded by extreme poverty, poor housing, health and sanitation, low literacy rates, limited access to support services, resource opportunities and social exclusion. Women that engage in fry collection and other fishery related activities usually do so because they have no other viable alternative. The process of uncontrolled population growth has reduced the per capita cultivable land and created overcrowding and high rate of disguised unemployment.

In view of existence of productive agro-climatic and soil morphological conditions and unique estuarine river system, a productive fishery resource has been emerged in the state of West Bengal, India. With its limited resources of men and materials, the Department of Fisheries (DOF), Government of West Bengal (GOWB) along with its functional organizations is making all out efforts in augmenting production by exploiting potential sources of pisciculture in all kinds of water resources including rivers, beels, baors, tanks, reservoirs, hill-streams, estuaries, sewage-fed water bodies, brackish water areas, inshore and off-shore areas and the deep seas. The entire frame of activities of fisheries development has been re-oriented with bulk emphasis on promotion of pisciculture by ensuring people's participation on a larger scale in the pisciculture activities. State inland aquaculture has emerged as first growing enterprise and viable alternative to the declining capture fisheries during the last few decades. The fisheries sector plays a very vital role in the state economy. More than 2.4 million fishers and other people engaged in this sector. The State of West Bengal has 8.21 per cent of total water resources of our country. It is always at the top in terms of total fish (18.36 per cent of the country) and seed production (55-57 per cent of the countries total seed production). In inland sector about 30 per cent of all India fish production comes from West Bengal. The fish production has increased from 2.73 lakh tonnes in 1982–83 to above 14.46 tonnes in 2007-08. Even if we consider the State Domestic Products (SDP) fishery contributes 3.12 per cent in the year 2007-08 which occupies the 2nd place in the state among the agriculture and other activities. The state is National Productivity Award winner continuously for last fifteen consecutive years for fish production. Ornamental fishery has a bright prospect not only in some domestic market but also has tremendous export potential as well. The state is pioneer in ornamental fish culture. The state is no. one in export of ornamental fish and has maximum share (70 per cent) in internal market distribution. The state is the highest in human resource exploitation and in women self help group involvement in ornamental fishery sector. About thirty thousand women fishers are involved in this sector. More than

132 such women groups are running successfully in this state, which is a unique involvement in the state of West Bengal as well as in India. In terms of recovery of loan amount, women SHGs performed better than male groups.

8.1. Role of DOF, GOWB In Ornamental Fishery Sector

- ☆ To explore the possibility of ornamental fish culture, DOF, GOWB has introduced the programme through the National Co-operative Development Corporation (NCDC), 'Rahtriya Samabay Vikash Yojona' (RSVY) and Fish Farmers Development Agency (FFDA) schemes.
- ☆ DOF, GOWB with the funding of NCDC aid through West Bengal State Fisherman's Co-operative Federation Ltd. (BENFISH) has organized training and inputs on ornamental fish breeding and culture management and allied activities for the ornamental fishery women co-operatives members.

- So far 132 no of such active functional women cooperative group, each with 10 to 20 no of members have been promoted through training and supplied with the inputs.
- Training and input on ornamental fish trade was also given to 500 no of women in 2006-07 and 500 no of women in 2007-08 from South 24 parganas and Birbhum district with the aids of RSVY funding which is a unique step towards market organization in India.
- Research projects have been undertaken on gut content analysis of ornamental fish, detailed survey of fishery resources in hill region and breeding of endangered indigenous ornamental fish. Research on algae culture and live feed culture of ornamental fish are also running successfully.
- Books (no. 3) were published on Ornamental endangered Fish and fishery.
- Research initiative was made on Ornamental freshwater prawns in aquarium.
- A unique drug "SWASTI" has been developed at Aquatic Resource Health Management Centre (ARHMC) at Pailan to combat disease problem and health management of ornamental fish.
- Certificate course on Ornamental Fish and Fishery for 6 months has been conducted in collaboration with the Netaji Subhas Open University (NSOU) and Dept. of Fishery, GOWB.

8.2. Unique Steps in Multi Species and Multidimensional Concept for Ornamental Fish Promotion

- Naihati Hatchery in North 24 Parganas district of West Bengal has found out a place for itself in the fishery map of India in terms of food fish seed production.
- DOF, GOWB has taken initiative to introduce the concept of Multidimensional Hatchery among the Hatchery owners and today most of the hatchery owners are producing endangered species and ornamental species in the same fish food hatchery.
- Tinfoil barb, Gold fish, Red tail shark, Albino shark, Pacu, *Puntius* sp. Barbs, Aquarium sharks (*Pangasius* sp) and many other species are available here and regular breeding of the said species are observed in this belt.
- There are around 100 hatcheries in the said area and around 15 per cent - 20 per cent of them are involved in multidimensional approach of hatchery operation producing ornamental fish of both exotic and indigenous types.
- The major problem in this belt is the lack of micro credit and market link during export of ornamental fish seeds. The seed producers only supplying their produce to feed the internal market. As a result, in spite of producing good quality produce our country is not being able to make a mark in the trade of world ornamental fishery where its contribution is very poor (0.001 per cent).

8.3. Ornamental Fish Trade Scenario

There are about 694 finfish species living in the state water against 2200 species of the country as a whole.

There are about 400 varieties of ornamental fish of significant importance in comparison to only 20 no of highly important commercial varieties of food fish.

Ornamental fish culture has been in practice in some districts of the state. Ornamental fishery has a bright prospect not only in some domestic market but also has tremendous export potential as well. To explore the possibility of ornamental fish culture, DOF has introduced this programme under F.F.D.A and NCDC programmes in the districts of North 24 Parganas and South 24 Parganas. Upto 2007-08, 852 nos. of ornamental fishery units has been developed. The basic features of ornamental fish trade in the state is briefed as follows:

- ☆ There is a weekly wholesale market of Ornamental fish at Galib Street in Kolkata, West Bengal.
- ☆ Most of the fish sold here are of exotic variety because they have the highest popularity.
- ☆ Half of the total markets deals with Gold fish followed by Angel, Molly, Gourami, Guppy, Barbs and Tetras.
- ☆ The trade of these fishes is mostly capture based and at present these fishes have a steady market demand both in India and abroad.
- ☆ Production takes place through captive breeding mostly at Howrah district.
- ☆ Native ornamental fishes are also traded in the market like – Indian glass fish, Guntea loach, Argus, Tangra, Giant danio, Queen loach, Yoyo loach, Asian knife fish, Banded gourami, Dwarf gourami, Honey gourami, Stripped spiny eel Gangetic mud eel, Zebra danio, Pabda, etc.
- ☆ Some expensive fishes like Discus, Silver dollar, Black ghost knife fish, Oscar, Mono angel, Elephant nose, Flowerhorn, Firemouth, Piranha, Arowana etc were imported from other countries.
- ☆ Some indigenous species like Indian glassfish, Pabda etc. are exported to Thailand and Singapore where they were artificially colored and imported back to cater Indian ornamental market at a very high price.
- ☆ There exists a huge potential to utilize these indigenous ornamental ichthyofaunal resources of the state to capture the global market.
- ☆ Large consignments are shipped from this state, specially from Kolkata and numerous people are involved in this trade in various pockets *viz.* Howrah, South 24 Parganas, Uttar Dinajpur, Siliguri of which Howrah happens to dominate the trade.
- ☆ In view to empower women folks and erase the gender bias the Dept.of Fisheries, GOWB has encouraged formation of women fishery groups who have been trained on ornamental fishery.

8.4. Natural Habitat of Ornamental Fish Resource: The Opportunities

The northern part of the state of West Bengal, particularly the Himalayan range has been bestowed with a great wealth of perennial rivers and small streams. These rivers and tributaries, namely, the Teesta, Rangeet, Ralli, Rishi, Reyang, Gelkhola Pala, Mahananda, Balasun and others are sources of important sport and ornamental fish species, crabs and prawns which have high export value. These precious species have never been investigated properly nor have they been identified for commercial purposes. So far no research has been carried out on biological productivity of these upland rivers holding cold-water biotic community. There is no substantial data to come up with credible information to provide a firm base for formulating suitable guidelines for conservation, propagation and judicious exploitation of the existing resources. The major contribution of the state of West Bengal is 90 per cent in India's ornamental fish export. However 95 per cent of India's export of ornamental fishes is based on wild collection of fish from northeast and eastern states of India.

The state of West Bengal has a rich indigenous fauna of ornamental fish species especially from rivers of hill regions. Of all exports from the state 80 per cent are wild catches from the Terai region (Mukherjee and Sarkar 2005). A large number of indigenous fish has tremendous possibilities to include in the list for export as ornamental fishes. Many of the indigenous fish species enjoy a high export value. Unfortunately, the whole of the export requirement of these indigenous fish species is met through harvest from natural habitats (Plate 1-7). If properly these fish resources are exploited, it will not only indicate as the geographically indicated place of availability of the species but also help to catalyze the upliftment of socio-economic status of the fishers of the uplands areas of the state. Consequently, overexploitation has adversely affected their abundance in nature. Appropriate measures could be encouraged for development of artificial propagation of such identified species so that economic benefits can be derived without affecting their natural abundance. But the unawareness of the fishers of this area, it is still in the very nascent stage of exploitation. Whatever little quantity is being exported that only under the individual exporter's involvement. Very recently the DOF, GOWB has started taking steps for the development of this sector, however there is still number of lacuna to meet up the vast demand in the export market. In order to address this issue an endeavors has been made with an objective to list of those fish species, aware the fish catcher about those species and to teach them the related technology especially in the hill region of the state.

8.5. A Unique Story of Export and Exploitation

Out of 132 women Cooperatives in our state most of them have been brought under NCDC assistance. The Department of Fisheries has been able to construct tank infrastructure for women Ornamental Cooperatives, which they are using regularly. It has been observed that enhancing production of ornamental fishes indefinitely doesn't help the business because marketing ornamental fishes is still not a smooth affair in this state due to:

1. Lack of internal demand
2. Lack of publicity/awareness
3. Legal problems related to air transportation
4. Lack of organized Marketing system in this state

To draw equilibrium in their earnings the Ornamental Cooperatives also rear table fish's seed in their tanks at comfortable intervals. The introduced concept gradually gained success and Ornamental Cooperatives started producing table fish's seed along with ornamental fishes. West Bengal is therefore propelled into a position where it is the only state in India where table fish seeds are cultivated along with ornamental fishes. Of all exports from the state 80 per cent are wild catches from the Terai region mainly Jalpaiguri district, adjacent to Assam boarder and Barobhisa area, Cooch Behar and Dinajpur districts.

The DOF, GOWB has made a study during 2007-2008 in a tribal village inhabited by 'Rava' tribe who collects wild specimens using various sorts of traditional gears (Plates 8–17). The 'Rava Basti', (Plates 18–19) at Barovisa village is situated beside the bank of the river 'Jorai' near Silbangla Forest in Jalpaiguri district. Sampling survey was conducted in some pockets of the said area where native ornamental fishes are available. The pockets are freshwater zone and coldwater Zone of Jorai River. The specimens were collected using various types of fishing methods such as gill netting, drag netting, cast netting, bag netting and other local fishing contrivances. Samples were preserved in 8 per cent formalin and kept for identification. In next phase they were transported in live condition and maintained in the aquarium of the Aquatic Resource Health Management Centre, Govt. of West Bengal. Observations were carried out in respect of their adaptability in captive water, feeding habit, breeding procedure, size range, compatibility with other species, trading pattern, source of finance, micro enterprise development policy. The study was based on the surveys conducted in houses of tribal people. Both formal survey and non-formal survey with Participatory Rural Appraisal tools like interactive discussion, semi-structured interviewing, sharing of information and ideas were carried out following Anon. (1992), Townsley (1993) and Schonhut and Kievelitz (1994). Data on various aspects were also collected from 'Gram Panchayat' and related Departments. The information obtained from fishers pertained to crafts and gears used during catch, technical know how, fishing requisites, infrastructure availability, trade and market information etc.

8.6. Fish Species with Ornamental and Sport Fishing Value

The study reveled that about 30 species of ornamental fishes are available throughout the year in the Jorai river. Among them only 10 to 15 species have market demand. List of common fish species (Plates 20–38) of the rivers of North Bengal are depicted in Table 114. Recently the 'Chang' fish is very much available from these rivers and it has higher export value. But the catchers cannot get exchange-price properly. Earlier, the availability of the quantity of 'Sabuj chingri' and 'Rajabam/ Bamos' was much more but of late, its availability became lower than that of earlier.

For example, earlier the availability of fishes per netting was @ 1000 pieces where as now a days it reaches @ 10 to 20 pieces. This may be due to spreading of poison like 'Thiodon, Copper sulfate' etc. by some dishonest fishers on the surface of the river water to kill the freshwater edible fishes for fulfilling their misdeed ((Mukherjee, 2002)). As a result the quantities of the ornamental fishes are becoming less even if some species of ornamental fishes are losing its extinct.

Table 114: List of Common Fish Species of the Rivers of North Bengal

Sl.No.	*Common Name*	*Scientific Name*
1.	Garela	*Acanthocobitis botia*
2.	Loach redtail.	*Aleorichthys sp.*
3.	Punti	*Barbus ticto*
4.	Jewelled chameleon.	*Badis badis*
5.	Punti	*Barbus conchonius*
6.	Melon barbs.	*Barbus melanampyx L*
7.	Tiger loach.	*Botia lohachata M*
8.	Giant gourami wild.	*Colisa fasciala*
9.	Point headed gowrami.	*Ctenops nobilis L*
10.	Chepti	*Cyprinion semiplotus*
11.	Bhitti/Punti/Naptee	*Danio aquipinnatus*
12.	Bhitti/Punti/Naptee	*Danio dangila*
13.	Bhitti/Punti/Naptee	*Danio devario*
14.	Big headed clown cat.	*Gagata viridescens*
15.	Blue striped garra.	*Garra apecies*
16.	Red tail garra.	*Garra sp.*
17.	Banded goley.	*Gobiodus sp.*
18.	Polack shark.	*L. calbeasu S*
19.	Shrimp red claw.	*M. peguensis*
20.	Shrimp chameleon.	*Machrobrachium scabriculs*
21.	White gloven shrimp.	*Macrobrachium sp.*
22.	Leaf fish.	*Nandus nandus*
23.	Fire barbs.	*P. conchonius (var) L*
24.	Shrimp rainbow.	*Palaemon henderso*
25.	Mudskipper (New).	*Periopthlmus sp.*
26.	Kaprey	*Pseudocheneius sulcatus*
27.	Shark roseline.	*Puntius denisonii*
28.	Dwarf golden barb.	*Puntius gelius L*
29.	Pipe fish (F.W.).	*Synguathus spicifer*
30.	Lyretall Red eye coral.	*Xip [Hophorus sp.]*

8.7. Livelihood Pattern of Rava Fishers

Many communities are living in 'Rava Basti' but most of them are the tribal people. The Ravas depend on the natural resources lying within their reach. The system of control and exploitation of land, water, and forest resources, the indigenous labour system, system of marriage by service, concept of totemism gradually going into oblivion, marriage celebrated at the boy's house, the belief that previously only the daughters inherited the ancestral property, the matrilineal descent system, importance attached to the matrikin, the aspect of filation and such other cultural traits make the Ravas important from anthropological stand point. Befor their Hinduisation the tribal priest (Deosi) of the Ravas used to perform worship *viz. Nagu Kan Bakau* in the month of April, *Mahakal, Thansiri, Michil Bay tanga*. They performed *Matri Puja, Nur Tanga or Akash Kamini Puja* in case of child birth. However, these tribal people are facing various types of acute problems from different aspects. They are so poor and ignorant that they can not receive their basic requirements of their survival. They have no alternative way to get rid of these problems. Though they are mainly depending on the agriculture for earning but this agricultural-job is a seasonal work. So they are facing disguised unemployment in the agriculture base work. There is no wide scope for earning from the agricultural base work, which is a seasonal work. Naturally they will have to find other alternative source of income to maintain their livelihood. So the other main source of income of these people is to catch ornamental fishes from the adjacent rivers namely Jorai, Mansai and Raidak. They used to go for community fishing during season. At the time of community fishing (Bahamara) some four to five hundred people gather from the near by villages. In response to the sound of bugle of buffalo horn, they shout saying "*Buro thakur bole Hari Hari Bol*" and after that only they start fishing in the river Rahidak which was previously known as Buranadi (Raychaudhuri, 2007).

Mainly the female members as well the children are engaged in collection of ornamental fishes and the male members are engaged to sell these collected fishes to the middlemen (broker) who are the main mischief-maker to deprive the catcher from their bonafied earning as because the fish catcher do not know exactly where and to whom their catch to be sent and sold. For this reason these innocent catchers will have to contact to the middlemen to sell their collecting fishes and in a single word, the fish catchers will have to surrender their collecting fishes to the middlemen with a very negligible price. As a result the original fish catchers earn a little money whereas the middlemen earn huge money. The fish catchers usually sell the ornamental fish @ 0.07 to 0.1 INR/piece whereas the middlemen sell the same fishes after purchasing from the catchers @ INR 0.60 to INR 1.50/piece to the exporter. It is noted here that the exporter's selling price is totally unreported because there is no concerned authority that regularly look after about the actual price by which the fishes would be purchasing or selling. No Govt. intervention has been noticed in this aspect when the volume of money infiltrating due to this system runs in millions.

8.8. Fish Species with Ornamental and Sport Fishing Value

There are more than sixty indigenous fish species in the hills of Northern West Bengal. The chief species encompass the prized fishes like "Asala" or Snow Trout

Plates 167–170: Traditional Gears Used at Rava Village

Plate 167

Plate 168

Plate 169

Plate 170

Plate 171: Insects Used as Bait during Fish Catch by Rava Fisherwomen

Plates 172–173: Rava Basti at North Bengal

Plate 172

Plate 173

Plates 174–177: Different Types of Local Fish with Ornamental Value at Rava Village

Plate 174

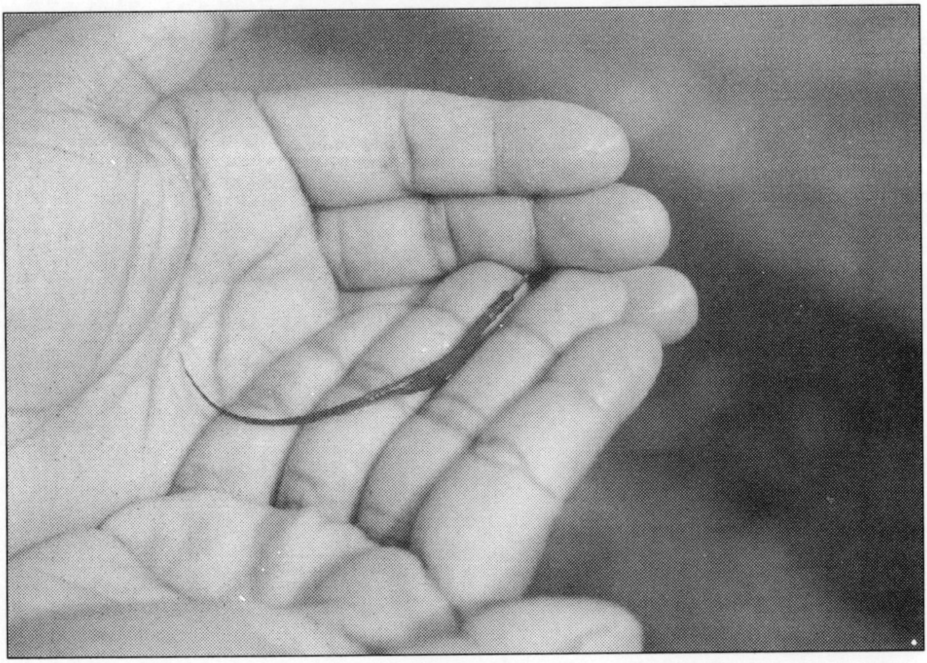

Plate 175

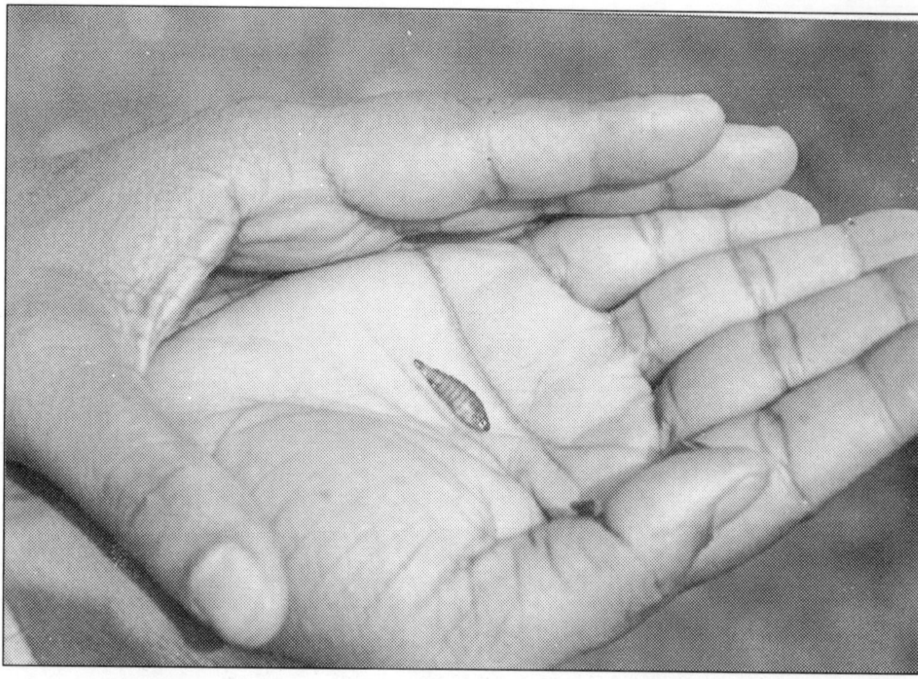

Plate 176

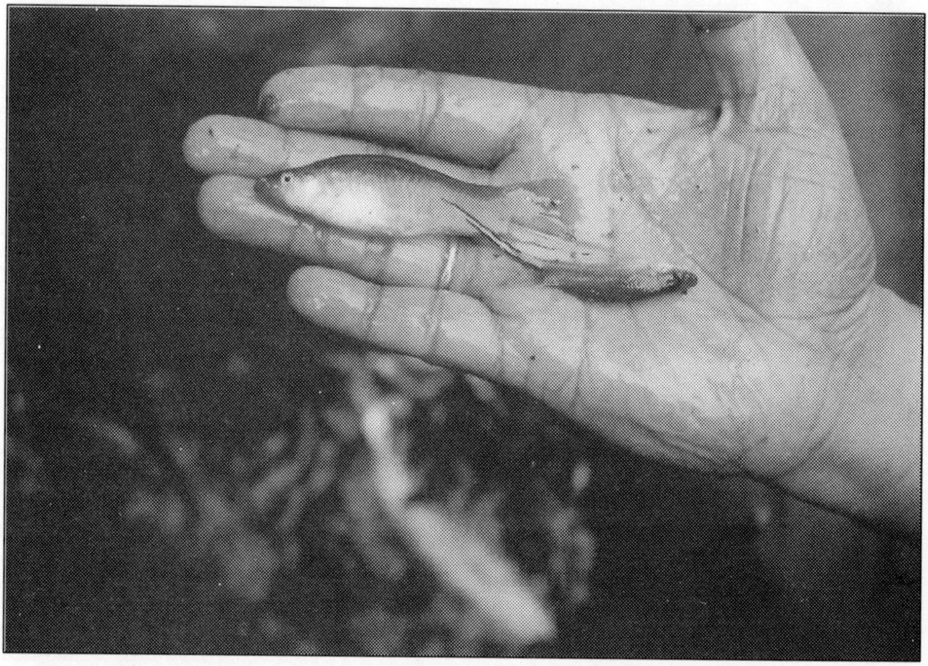

Plate 177

(*Schizothoraicthys* sp.), "Katli" or Copper Mahseer (*Neolissochilus hexagonolepis*) and the "Mahseers" (*Tor* Sp.). These three mentioned are highly valued considering their table and sport fish values. The biology, behavior and propagation of *Tor* species are well known and understood at a number of Government run breeding centers around the country (Rao, 2006).

(*i*) Barbs. Carps and Minnows

The important family Cyprinidae comprises a very large and dominant group of primary freshwater fishes. Given below are brief descriptions of each group relevant to the Hills of Darjeeling.

1. The subfamily schizothoracinae which includes Snow Trouts is present in the Himalayan and sub-Himalayan regions of the Indian subcontinent. Snow Trout prefers rapids and pools of large torrential streams fed by snow-melt with water temperature from 8° C to 22° C. There are around six species found here.
2. Copper Mahseers (*Neolissochilus hexagonolepis*) is an important sport fish. In India this fish is known from North Bengal under its Nepalese name "katli", and the fish is also present in Assam and Myanmar and in the River Cauvery in Tamil Nadu. These fish have a preference for water temperature of 10° C–30° C.
3. Mahseer or *Tor* is the most important sport fish of India, Pakistan and Nepal. Of the seven species of *Tor*, only *Tor tor* and *Tor putitora* are reported from Bengal. These have a wide distribution in Himalayan Rivers from 135 m to 1650 m altitude.
4. Cyprinion is mountain stream cyprinids that are characterized primarily by the possession of sharp, horny edge to the lower jaw and no fleshy lower lip. The distinctiveness of these fishes lies in the size and shape of the head and extreme body depth.
5. *Labeo* is very widely distributed in South and Southeast Asia and Africa. The genus is very well represented in India where some of the species reach a length of 1.5 m. Because of their varied habitats and also extensive cultivation, interbreeding has taken place giving rise to several color forms, varieties etc. Even inter-specific and intergeneric crosses are known.
6. *Puntius* form a great group of Barbs which are widely distributed in South Asia and are represented in the Hills by a few species.
7. Hill Trouts or Barils of the genus Barilius are for the most part fishes of hill or mountain streams, although some species inhabit lowland waters. These fish are silvery, usually with dark spots or bands on the body. Many species are beautifully marked.
8. *Danios, Devario* and *Danionella* are small, slim, attractively colored and active. These inhabit small waters and are commonest in mountain rivulets.
9. Algaeaters or Stone rollers are consistent with members of the genus *Crossocheilus* and *Garra* which dominate fast flowing waters between 1250 m -1875 m. The outstanding diagnostic character of this group of fishes is

the absence of a groove between the upper lip and the snout, the upper lip being coalescent with the skin of the snout to form a sucking disc. This sucking disc or sucker mouth is an ideal adaptation as it allows the fish to attach themselves to rocks or stones and remain in the current while feeding giving rise to their popular name "stone rollers". In addition many have quite large pelvic and ventral fins, which assist in holding the fish in place while grazing on algae and microscopic life.

10. *Psilorhynchus* species are known to occur primarily in the Gangetic drainage. This family comprises a monotypic genus only of small hill stream species. A fairly common fish, *Psilorhynchus balitora* is found in the fast streams and shallow rivers, especially where the bottom is rocky, adhering tightly to the rocky substratum with the expanded paired fins and the breast applied to the rocks. Like other torrential fishes, these always point their heads against the flow of current. The variegated coloration of this fish harmonizes so perfectly with the surroundings that it is hard to detect its presence even from a short distance.

(*ii*) Hill Stream Loaches

1. The Balitorids or members of the Balitorinae family are small freshwater fishes distributed extensively in the hills of Southeastern Asia. These generally have rounded heads and streamlined bodies, adapted to life on the bottom in very swift mountain streams. Their paired fins act as adhesive organ. These fish attain a maximum length of 12 cm.
2. Members of the subfamily Nemacheilinae are popularly known as Stone Loaches. These loaches are mainly found in mountain and submountain regions but rarely in the plains. They live in shallow, clear and swift streams among the stones and pebbles at the bottom. Environmental conditions tend to produce similar characters and it is probable that the great resemblance between the various species of nemachilids is due to a similarity in their environment.
3. Sting loaches of the family Cobitidae are mostly small, elongated and anguliform bottom dwelling fishes occurring under pebbles or sand in swift streams, rivers and some lakes. These loaches are generally species of small size that range in body size from worm-like or chunky bodies such as certain species of *Botia*. The Lepidocephalids are small fishes found mostly in swift streams but also in swamps and lakes. They appear to prefer sandy or fine gravelly bottom, into which they are able to burrow and quickly disappear, if alarmed. The spinous first pectoral ray helps in "digging in". The botias are in principle inhabitants of running waters, mostly in upland areas. The large bifid suborbital spine which can be erected and set in a position perpendicular to the cheek, must be of material value in discouraging attacks by snakes and fish. Some what, it is similar to the tusks of a wildboar.

(*iii*) Catfishes

1. Torrent Catfishes are members of the genus *Amblyceps* which are found to inhabit swift streams. *Amblyceps mangous* a small catfish not more than 7 cm in length is very similar to look like a Magur (*Clarias*) or a Singhi (*Heteropneustes*) but the distinguishing feature is its emarginated or forked tail.
2. Sisorid Catfishes belong to the exclusively Asian family Sisoridae which are bottom-dwelling catfishes most with more or less thickened leathery skins. They typically inhabit high gradient low land or mountain streams with adults ranging in size from 20 mm to 2 meters. Many species have adhesive organs (labial or thoracic) with which they cling to the substrate.

   Sisorids comprise many genera and notable among them are *Bagarius, Conta, Coraglanis, Erethistes, Erethistoides, Euchiloglanis, Gagata, Nangra, Glyptothorax, Hara, Pseudalaguvia, Myersglanis, Laguvia* and *Pseudecheneis.*

(*iv*) Freshwater Eels

Anguillids live in a variety of freshwater habitats such as streams, rivers, ponds and lakes and are catadromous. Spawning occurs in the open ocean, and after drifting in the surface waters for several weeks, the larvae called leptocephali are carried near shore where they metamorphose to the tiny, transparent glass eel stage and enter rivers. Once in freshwater, the glass eel becomes pigmented and is called an elver, the elvers migrate up rivers. After 10 to 15 years in freshwater, the eels migrate sea as adults to spawn offshore.

8.9. Coldwater Fish Potential for Ornamental Fish Exports

Darjeeling Hills is bordered in the west and north west by Nepal, in the north east by Sikkim and the east by Bhutan, where snow-clad Himalayas, eternal glaciers, ice cold torrents, clear waters contribute to much of its hydrosphere. These vast stretches of inland water support many and varied forms of freshwater life including fish. The need and opportunities for the study of the freshwater bioresource and its management are enormous.

Many wild and scenic rivers and lakes of these hills provide opportunities for recreational fishing, flow raising and aquaculture. More progressive, dynamic and imaginative management techniques are to be developed for sustaining harvest of fish from these waters and integration of aquaculture. From gazing at the azure blue waters of hill streams and rivers did the pioneering idea germinate to utilize a part of the coldwater fish wealth for the Ornamental Export trade and thus earn valuable foeign exchange for our State and Country (Rao, 2006).

8.10. Water Resources

The Teesta, the most magnificent river of Darjeeling District, has its source in Lake Cholamo, situated at an elevation of 17,500 feet, which exceeds twice the height of the well know Tiger Hill of Darjeeling by 500 feet. This lake lies to the north of the Donkia Pass near Shetschen, where from the summit of the pass is almost 8 Km away to the north-east of Darjeeling as the crow flies.

On its way, contributions are made to it by streams which rise in Thangu, Yeumthang and Donkia Laranges far, far away from and due north of Gangtok, the capital of Sikkim. Then, as it approaches the Teesta Suspension Bridge, which joins Kalimpong with Darjeeling, it receives a mighty torrent of later from a stream called the Great Rangeet. From this confluence of the two rivers, the aforesaid bridge is 7 Km. The Teesta river debouches through the gorge at the Teesta Bridge, at an almost imperceptible gradient, taking a southerly course and finally enters into the plains through Sevoke, Nhere it has been spanned by the Coronation Bridge. It would be of much interest to many to mention here that this river passes through Jalpaiguri and the Rungpore Districts in Bangladesh and falls into the mighty Brahmaputra at Fulcherry. The combined waters then flow on to Gualundo, where the Padma, uniting with them forms the unbounded expanse of waters known as the Meghna river over 80 Km in width, the tall palm trees on either bank look like so many tiny plants or even dots.

The Great Rangeet takes its rise from the glacier of Kabru and proceeds southwards till it meets the Rammam river coming down its source near Phalut in the Singalila range. The combined waters, after "traversing just half a kilometer, take up a further tribute from the Little Rangeet hailing from the base of the Tonglu spur in the Singalila range, and these form the Great Rangeet which flows on for a distance of about 15 Km till it merges with the Teesta river.

The Mahananda river has its source near Mahaldivam Hill to the east of Kurseong. As soon as it debouches through a gorge of the Himalaya, it runs through a subterranean stratum of sand for a distance of nearly 7 Km before heaving into sight again. In the rainy season, this queer phenomenon disappears with the rushing in of a mighty torrent from the hills, which makes the river overflow its banks. This river has been spanned by an iron bridge to the north of the town of Siliguri. At a distance of 7 Km from the bridge and in the downward course of the river, it receives its contribution from the Balasun river which flows part of the valley below Kurseong (Rao, 2006).

The other rivers can be listed as the Mechi, Reshi, Ghish, Chel, Murty, Diana, Jaldhaka, Torsa, Rydak and II and Sankosh among many other rivers and numerous hill streams. Mention should also be made here of the Mirik lake which is home to a couple of endemic fish species.

8.11. Rampage of Ecology and Destruction of Fauna

Ecology of the hill streams in general have considerably been deteriorated due to heavy land slides since 1968 resulting in the rise of river beds, disappearance of a good numbers of permanent pools harbouring game fishes. Despite of this rampage of nature they still hold enough prospects for development under planned programme and conservation. Fishes-young, adult and brood are found naturally trapped in stagnant transparent pools bounded by boulders (created by the river itself) often met their ruthless destruction by explosives at the hands of merry goers of picnic party / poachers.

Impact on human intervention on aquatic ecosystems and eventually on the native coldwater fish community is viewed as a serious problem. As fishing with nets is considered by many as a slow laborious process often with very poor catches, people have resorted to "unfair" practices to catch fish which is detrimental to the environment as a whole. In the upland section of the rivers practice of electro fishing and tossing of dynamite into water is in vogue. Rivers running parallel to Tea gardens are poisoned frequently with pesticides such as Thiodine used in the Tea industry. Such destructive activities should be stopped (Rao, 2006). "Dealipara" is a process of wanton destruction of fish life in small streams by taking recourse to diversion of the flow of streams from its normal course thereby drying up the prospective pools having plenty of fish. Killing of fish with pesticides, copper sulphate, bleaching powder is also prevalent. Nowadays the cheap availability of pesticides in this area allures the delinquent to take recourse to such procedure for mass killing of fish. Often some quantity of copper sulphate tied in a piece of cloth is hung over a prospective pool to kill fishes. Persons staying in remote rural areas use fruits of 'maidal' plant (*Erythrina arborescens*) or barks of "arare" creepers (*Acacia concinna*) to kill fishes in streams and rivulets. All these types of wanton destruction of fish life are the main factors that can be attributed towards depletion of fish in hill streams.

8.12. Coldwater Aquaria

There has been in existence from long in European countries, USA, Canada, Japan and China the Ornamental Coldwater Aquaria. For centuries, China and Japan have perfected strains of their Goldfish and Koi giving rise to a wide choice for the hobbyist. All these fish are raised in farms and distributed worldwide. Recently the trend has begun to introduce various coldwater species for use in aquaria (Rao, 2006).

8.13. Constraints Faced by Local Fishers

- Excess sediment particularly during the rainy season increase turbidity of water resulting in decrease of food organism of the system.
- Soil erosion following deforestation results in siltation covering spawning sites with sediments that can ultimately become a cause of loss of fish diversity.
- Dams act as physical barriers for migration, preventing access to fishes to their usual breeding, rearing and feeding grounds.
- Indiscriminate use of nylon gill nets of small mesh size, attempt to catch all the fishes in the dry season, to catch fishes from breeding ground results in severe population disturbance.
- Killing of fish by pesticides (Chlorinated hydrocarbons, organophosphates and carbonates), copper sulphates, bleaching powder are rampant.
- The ornamental fish catchers have been depriving from their actual income due to the interference of the so-called middlemen.
- Non-availability of crafts and gears in time and space.

- ☆ No electricity availability, so establishment of hatchery could be rarely possible.
- ☆ No rhythmical availability of the quantity of fishes throughout the year.
- ☆ Lacking of proper training regarding culture, maintenance and transport of their produce.
- ☆ Communication gap between the exporter and the catcher.
- ☆ Non-availability of oxygen cylinder for maintaining their catch properly. They used to buy each pouch of Oxygen @ INR 5/-.
- ☆ Lack of budgetary support.
- ☆ Non availability of hatchery or seed bank
- ☆ Poor infrastructure and undeveloped market linkage system.

Chapter 9
Scope for Game Fish

The chief interest of a visitor (especially foreigners) to this area, however, lies in the sport fishing that is provided by certain well-known game fishes. All the hill sections of the largest streams and especially, the Teesta River abound in Mahaseer, Katli, Indian trout and Goonch, generally fished, in clear running streams by means of rod and lines. The junctions of a smaller stream with the main river are usually most suitable for angling. The premier place among Indian game fishes is occupied by the Mahaseer–a popular name for varieties of large scaled carp. The portion of the main stream, Teesta, touching its border from Sikkim is inhabited by various small fishes having cylindrical shape and organ for attachment to the rocks to cope up with the fast current of the streams. These fishes are small in size having low nutritive value, slow growth and as such, may be not of commercial importance but having immense ornamental potentiality from the export point of view. In the next phase of the stream down the plateau where its course has been wider and velocity comparatively slower due to less rapid rise and fall of the rivers, offers scope for culture of commercial important species *viz.* Katli, Mahaseer, Gurdi, Indian snow trout etc.

The pleasure in Mahaseer angling is that its endless thrills and excitement to the anglers. It is great game fish and very courageous, fighting to its last gasp and splash. Once bitten by mahaseer fever, anglers will never hanker for the other fish. As back as 1926, it is on record that, James Edward Corbett, in the spring (March, 1926) angled two massive Mahaseer, 30 pound and 35 pound respectively to the envy of the hundreds of local hill people of Rudraprayag. It is recorded that the locals survived much upon the abundant shoal of Mahaseer, Gooch and many other coldwater species, which thrive in the cold regions of the country.

It is one of the world's finest sporting fish on which classics like 'Circumventing the Mahaseer' (Macdonald) with rod and line in India. The mighty mahaseer

(Skensdhu) The complete Indian Angles (Master), Sunlit waters (Conway), to name only a few have been written by British authorities. Anglers from all over the word have had the thrill of circumventing this noble fish in our country; there has been no dearth of enthusiastic and successful anglers.

The Katli (*Neolissocheilus hexagonalepis*) is another sport fish of the hills, possess large scales and in general appearances similar to copper or chocolate mahaseer, grows upto 5.5 kg or 60 cm in length. The Katli seems to be a very suitable fish for culture both on account of the ease with which it could be bred and its omnivorous feeding habits. This species is recommended for culture in artificial pools fed by perennial Jhoras (springs).

If the dynamiters of the tributaries and off-streams can be put a stop, then nature herself could be able to replenish the stock of Mahaseer. The Mahaseer would again team in countless numbers in the Teesta and her tributaries making the Teesta basin truly the paradise for anglers, when the mahaseer would reign supreme once more, as king of our mountain rivers bringing thrills and excitement galores to the sporting fraternity of anglers.

The mahaseer used to teem in the Teesta river and its tributaries in the past. They were then caught by hooks and lines only. In those days the Teesta and its tributaries flowed through green, virgin, forests. The river banks were devoid of any erosion. Construction and felling of trees along the river banks were few and far between and the waters flowing from the upper reaches were limpid, crystal clear and as pure as the snows of Kanchanjunga (the sources of the Teesta and Rangeet rivers). Fish teemed in them in thousands. Nature smiled and the Teesta was indeed an anglers' paradise. Mahaseer swam there in thousands in their natural habitat in the big pools of the Teesta where the water ran deep and blue. It was quite common then to have mahaseer six to eight feet in length and the fish could actually be seen splashing in river with huge record sizes dorsal fins.

It was only in the late twenties that the anglers of those days having discovered the thrills and excitement of mahaseer fishing decided to conserve and protect this wealth. A club was formed known as the Darjeeling fishing and shooting club. The club was formed mostly by the then British tea and cinchona planters, some members from the railways and a few local enthusiasts. In those days, well known fishing spots were the confluence of the Relli-Teestas, Reyang-Teesta and Rangit-Teesta. The fishing and shooting club took lease of the fishing right from the Forest Department and at the above mentioned location the club stationed uniformed guards known as the fishing and shooting patrol who used to regularly patrol the rivers at these spots and arrest poachers. As explosive were rare, there were virtually no dynamite poachers.

Members of the club were also empowered to arrest all poachers. Any anglers or hunter when challenged by a member had to produce their shooting or fishing license. Those were the days of glorious fishing and well-known anglers like Mr. Richie, Mr. Samson Day, Mr. Langdale Smith, Mr. John Macdonald to mention a few, whose names will remain as pioneers of Mahaseer fishing in the Teesta till the dismantling of the Railway Inspection Bunglow at Reyang, visitors could see outside the Inspection Bunglow a plank measuring 4 ft. by 6 ft. on which was traced and outline of the record catch of gooch caught by Mr. Samson Day weighing 24 pound while inside one of the rooms the wooden paneling was adorned with outline tracings of mahaseer caught by Mr. Ritche. Even the fish scale of the catches was tacked on the figures and inscribed in these figures were the weights in pounds and the date of the catch.

Unfortunately the membership of the fishing and shooting club declined with the departure of the British planters, the dislocation of rail communication and restrictions on shooting game. Finally in the mid-sixty the club was dissolved and it becomes open for the poachers. Fortunately, as there were few constructions nearby, the use of explosive was restricted to a few blasting of pools in the smaller off stream. Poaching was done mainly with hooks and nets. The use of poison was very rare in those days. There were anglers but few in number and thus the fish continued to teem in the Teesta and its tributaries.

For efficient management of the rivers and lake, these should be opened to the anglers in order to stop poaching as well as the control the dynamics of fish population. Earlier in June 1985 the Mirik lake was opened to the anglers for a few days but has to be stopped for some obvious reasons.

For proper implementation of the programme, an advisory committee may be formed under the leadership of the Sub-Divisional Officer, Kurseong, Project Officer Tourism Mirik, Circles Inspector of Police, Kurseong and the Fishery Councilor, Darjeeling Gorkha Hill Council, Mirik. Broad guidelines for angling at Mirik Lake:

1. Every year April to June and September–October will be opened for angling.
2. Angling time will be from 10 a.m. to 3 p.m.
3. Only twenty numbers of angling permits will be issued per day @ Rs. 100/- each.
4. The bag limit per angler will be restricted to 5 kg. Any excess to that quantity will be charged @ Rs. 50/- per kg.
5. No angler will be permitted to catch individual fish below 750 gm.
6. The sale proceeds from the sale of angling permit/fish may be kept in a Bank account in the name of the Advisory Committee and may be utilised for the development of fishery in the lake
7. The angling may be checked by the officials authorised by the Advisory Committee.

Chapter 10

State Programme for Capacity Building on Aquaculture Development and Diversification

With the production of over 3.3 million tones from the inland fisheries sector, India has occupied the second position in the world. Such growth has been possible due to impressive development in the aquaculture, mainly through carp culture in freshwater ponds and tanks as also the integrated fish farming practices. During the past two decades, the inland aqaculture fish production has increased to 2.9 million tones, with carps alone contributes about 85 per cent. The state of West Bengal contributes about 37 per cent of all India Aquaculture production. The fish production has increased from 2.73 lakh tonnes in 1982–83 to above 14.46 lakh tonnes during the year 2007-08 of which the inland sector contributes 12.64 lakh tons. Even if we consider the State Domestic Products (SDP) fishery contributes 3.12 per cent in the year 2007-08 which occupies the 2nd place in the state among the agriculture and other activities. The state is National Productivity Award winner continuously for last fifteen consecutive years. Almost 89 per cent of the inland fish production comes from Aquaculture sector in the state. Appropriate technological interventions as well as extension services provided by the Department of Fisheries, GOWB in time and space has led to this success. Even after that, Aquaculture is further poised to reach new heights in terms of enhancing production, productivity and profitability.

With the augmentation of demand for human food supply to support its ever increasing population and also to support the livelihood of the community associates with it and improve their economical status, it was felt necessary to utilize the available limited resource in most optimal manner in order to extract maximum possible out put from the existing resource without affecting the normal balance of nature. This has led to intensification of aquaculture by improved technology with high stocking density, scientific monitoring with controlled manureing and supplement feeding and introduction of fast growing compatible species like Pangas, Paku etc. replacing or in combination with conventional culture of IMC and Exotic species. Introduction of Integrated fishery is also another such approach where Aquaculture practice is integrated with culture of other domains such as livestock and horticulture with the basic theme that the waste and byproducts of one culture forms the natural input resource of another leading to optimum resource utilization and giving maximum out put in terms of economy in most eco-friendly manner. Diversification of Aquaculture by culture of various organisms of aquatic origin to satisfy humans' need in various others ways mainly in terms of economics, in addition to food supply such as cage culture, mericulture, seaweed culture, pearl culture, oyster culture, ornamental fish culture etc. may also be taken up to achieve the similar goal as per availability of local resources.

10.1. Need for Diversification in Aquaculture

India is the second most populous country in the world, with a 2003 total population of 1.068 billion, representing 17 per cent of the world's population, occupying 2.4 per cent of the world's land mass. The overall annual growth rate has been registered at a rate of about 1.78 per cent per annum. It is expected to overtake that of China, which has thrice our geographical area, in another two decades. The state of West Bengal is the fourth most populas state situated in the Eastern Region in India. 82 million people stay in the State. It is 2.7 per cent of India's area but about 7.8 per cent of the country's population, this state ranks first in terms of density of 904 per sq. km as per the 2001 census. The proportion of the rural population in the state stands at 72 per cent. A significant part of the state is relatively more economically backward and also tends to be less advanced in terms of human development. The process of uncontrolled population growth has reduced the per capita cultivable land and created overcrowding and high rate of disguised unemployment.

Under this circumstance, the greatest challenge is to ensure the food security of the largely undernourished and protein starved populace in rural as well as urban area, especially in the context of declining land resources available for animal husbandry and agriculture due to urbanization. Hence Fisheries, mainly the Aquaculture sector will be the savior to meet the increased demand. Unlike agriculture and animal husbandry practices, fisheries activity does not consume water and make a significant demand into the already scared water resources. On the other hand, it can help to improve the quality of any open waters. In view of existence of productive agro-climatic and soil morphological conditions and unique estuarine river system, a productive fishery resource has been emerged in the state of West Bengal. With its limited resources of men and materials, the Department of Fisheries (DOF), Government of West Bengal (GOWB) along with its functional organizations is making all out efforts in augmenting production by exploiting potential sources of pisciculture in all kinds of water resources including rivers, beels, baors, tanks, reservoirs, hill-streams, estuaries, sewage-fed water bodies and brackish water areas.

Moreover, the net impact of global warming so far has been modest, but near-future effects are likely to become significantly negative, with large-scale extreme impacts possibly by the end of the century. The predicted effects for the environment and especially for aqua resources are numerous and varied. Climate changes have a lot of impact on fish biology. As water warms up, many parasites and microbes that cause fish diseases grow faster and become more virulent. Massive fish die-offs due to toxic algae and the risk of human illness from eating poisoned fish will also increase as temperatures climb. Global warming is thought to be an overall effect of anthropogenic pollution. For preservation, restoration and propagation of these species, pollution is to be checked. Microbiology and Parasitology unit of Fisheries Department, Govt. of West Bengal (GOWB) at Pailan for fish health management was established for researches on fish diseases and culture on enhanced technology for production of different varieties of fishes and aquatic species. The aim is to meet the increased demand of fish, along with judicial utilisation of water resources.

Increasing incidence of obesity coupled with stress, blood pressure, heart diseases and diabeties and awareness leading to health consciousness has now turned the attention of the urban populace from red meat and junk food to fish–a source of cheap protein, vitamins, iron, phosphorous, and especially omega 3 fatty acids. The rural poor people in most cases suffer from malnutrition of all kinds, especially the children. Keeping in view of the large yield gap to meet the minimum nutritional requirement of 11 kg per capita as against the present 9 kg through out the country, the XI five-year plan envisages almost doubling the aquaculture production. However this should be done in a judicious way, which is far more important than the short-term goal of realizing the present targets. Under this circumstance, it is needless to mention that in inland fisheries, the yield gaps are need to be filled in order to realize the full potentials by intensification, integration and diversification of aquaculture and allied activities in the state of West Bengal.

10.2. Perspective Plan for Diversification of Aquaculture

The State possesses potential resources in freshwater and brackish water sector. It has been possible to cover the cultivable water area to a considerable extent under limited fund provision and other constraints. Still there is huge area including derelict or semi–derelict areas, which can be either brought under productive utilization towards pisciculture for production of more fish. Approximately 21 per cent of available area of ponds/tanks, 49 per cent area in Beel and Boar, 81 per cent of area in Reservoir, 100 per cent area in Canal and 77 per cent area in the brackish waters, beside huge areas under borrowpits both on the highway side and railway lines, Jhora (hill streams) are yet to be tapped. Hence, it is envisaged to develop the available resources in all these sectors by adopting improved culture practices. There is also tremendous scope for augmenting the production by manipulating fish combinations of potential fish species. Though we are rich in resources, the water bodies can not be utilized properly until and unless the complete right of these resources goes to Fisheries Dept., because sole right is necessary to implement any necessary rule, regulation, act or any welfare scheme for protection of these water bodies. Moreover, strong linkages between research and development agencies (both in the State level as well as between

National and State level) are pre-requisites for technology development and implementation as it is basically a multidisciplinary activity. These measures will result bringing more areas of resources under productive utilisation and enhanced productivity, which ultimately will increase State Fish production to a great extent towards reaching a Fish Surplus State.

(*i*) Introduction of *Pangas* for Enhancing Fish Production

It is observed that the hatchery owners at Naihati belt of North 24 Parganas district are producing huge quantities of fish seeds. This place has made its renowned position in the Fishery map of India as seeds are distributed from this place to all over the country. The hatchery owners in this area have more than 30 years of experience. They have developed their own technology, which needs proper scientific back up before standardization. One of their new innovations is seed production of Thai Pangas (*Pangasius sutchi*) in the Multipurpose and Multidimensional hatchery. Thai Pangas a silurid catfish is endemic to Thailand, Vietnam, Laos and Democratic Kampuchia, where it is cultured on commercial basis in closed water bodies *i.e.* pens and cages. The fish, due to its high growth rate has proven to be a suitable candidate for both mono and poly culture with carp in pond and in net cages. Presently a huge quantity of Pangas seed is being produced by a number of carp hatchery owner at Naihati in North 24 Pargana district of West Bengal and seeds were distributed all over the country where it has a great demand. This species was introduced in the eastern part of India through Bangladesh from Thailand during 1994-95 without following the rules of introduction. It is a fast growing species attaining a maximum size of about 120–145 cm. A large number of fish farmer in India have been attracted towards the commercial culture of this species due to its remarkable growth rates as well as hardy nature. It closely resembles to our native species yellowtail Pangas (*Pangasius pangasius*) which is included as a threaten species in the 39 species identify to be disappearing from their natural habitat in West Bengal and are required to be breed artificially for restoration to their natural habitat. The fish due to its high growth rate has proven to be a suitable candidate for both mono and poly culture with carp in pond and in net culture cages. Both the silver stripped and yellowtail pangas are being adopted recently for culture along with Indian major carps.

Special Features of *Pangasius*

- ☆ It is a freshwater fish that live in tropical climate and prefer water temp of about 22°–26° C.
- ☆ Wild fish prefers omnivorous diet consists of crustaceans, other fish, and plant matter.
- ☆ In natural environment, the reproductive season started from May to June and lasts from October to November.
- ☆ This fish is easy to farm, easy to process and fast in reproduction.
- ☆ Wild fish prefers omnivorous diet consists of crustaceans, other fish, and plant matter.
- ☆ For cage culture–a mixture of trash fish, rice bran and broken rice (9:1:1) is commonly used as feed delivered at the rate of 2 per cent of the body weight/ day.

- ☆ Other feeds include waste vegetables and soya bean waste.
- ☆ Seeds fish are produced in both government and private hatcheries.
- ☆ It is observed that hatchery owners in Naihati area used boiled pea nuts as food for brooders because they think that addition of vegetable protein helps to increase fecundity instead of using animal protein that produce excess fat and ultimately reduces the fecundity.
- ☆ Some brooders are produced in cage operations, which rear the fish to a size of 3 to 5 Kgs in 2 to 3 years before they ripen.
- ☆ Spawning usually occurs in the period January to March.
- ☆ This fish has all benefits of a fantastic Aquaculture fish on its side:
    - It is growing fast – needs only 6 months to reach its weight of over 1.5 to 2 kilos
    - It is very robust, knows next to no diseases
    - It needs no high oxygen levels in the water.

The frozen fillets of Pangas have high export value.

The *Pangasius* comes in different colors from snow-white to light pink, pink and yellow. The white fish is the highest quality but light pink is requested the same in the markets.

Nowadays even so called "Bio Pangasius" is offered in the markets.

(*ii*) Introduction of Culture of Pacu, *Piaractus brachypomus* with Successful Breeding and Seed Production

Diversification of aquaculture is essential for the betterment of fish farmers in the present day situation. Keeping this view in mind Pacu, *Piaractus brachypomus* may be an important candidate species for mass production in aquaculture pond. Pacu, locally called Rupchanda is basically native of Amazon and Orinoco river system of South America. This fish is also known as red pomfret belongs to the family Serrasalmidae. This fish is a close relative to Piranha as the family is same. But differ in feeding habit and behavior. Rupchanda is basically a vegetarian fish but they can eat almost any food. It grows very fast and attends a weight of 1.5 kg in a year. It is a very tasty fish and hardy fish in nature. It can tolerate acidity and can also be farmed in neutral water condition. More over it can tolerate salinity as high as 10 ppt and dissolved oxygen as low as 0.5 ppm. (Datta and Nandeesha, 2006).

Apart from culture as table fish, Pacu has very high market demand as ornamental fish because it's beautiful look and presence in aquarium during juvenile stage. Study revealed that there is very rare occurrence of disease among Pacu. Though due to close resemblance with piranha Pacu is not advocated for culture in many countries due to fear of unwanted entry of the deadly Piranha fish in open as well as closed system, Govt. of Malaysia lifting the ban on culture of the fish. Department of Fisheries, Govt. of West Bengal like to promote the culture this fish species owing to its numerous positive side as mentioned above. For rapid promotion of culture of Pacu, standardization of seed production and larvae rearing technique is essential. So

Plate 178: A 5 kg. Female Pacu Brooder

Plate 179: Pacu Brooder Management Tank

breeding technique in captivity with hypophysation was successfully done by the Dept of Fisheries Govt of West Bengal.

Pacu, *Piaractus brachypomus* has a narrow plate shaped dark grey body, black upper body and back, and a small mouth with blunt teeth. Pacu possesses 31-34 gill rakers and about 85-94 lateral line scales on the gill arch. Pacu can attend 20 kg in size. For breeding purpose brooders may be collected locally and can be raised in the composite fish culture tank along with *Cyprinus carpio.* The size of the tank is 0.5 ha with a depth of 5 feet on an average.

Regular feeding should be done @ 6 per cent of the body weight of Pacu. Feed may be provided at the evening time just after sunset. Boiled peas along with fish meal and ground nut oil cake are used as supplementary feed. Aeration to be provided by means of pump. As the *Cyprinus carpio* is a good fish for bottom racking these fishes are liberated to utilize the excess feed provided to Pacu and for racking the bottom for better pond management. Brood fish were getting maturity for breeding after 3-4 years. For male 3.5 kg and for female 5 kg respectively is ideal weight for breeding. During breeding season sex can be easily identified by some external feature of the brood fish. Comparatively swollen belly with round reddish vent, female brood fish can be easily differentiate from male having pointed vent with depressed belly.

Brood fish may be collected from brooders tank one day before of breeding and kept separately sex-wise in hapa. No feed shold be provided during this period. Sex ratio breeding is 1:1. Pituitary gland extraction is used for hypophysation. Initial dose of hormone is @ 2 mg per kg body weight in female. No male are injected at the

Plate 180: Pacu Brooder

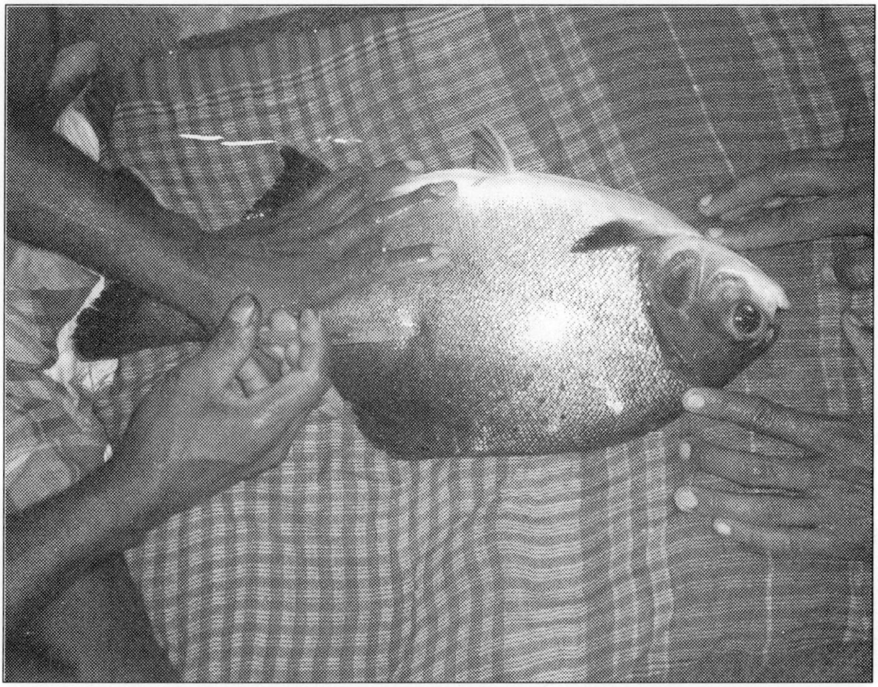

Plate 181: Pacu Brooder being Injected

Plate 182: Fertilized Eggs of Pacu

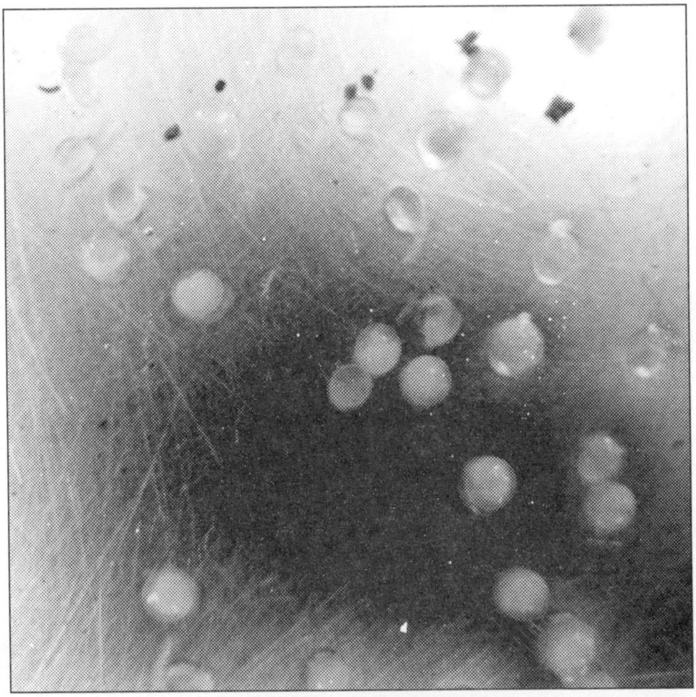

Plate 183: Just Hatching

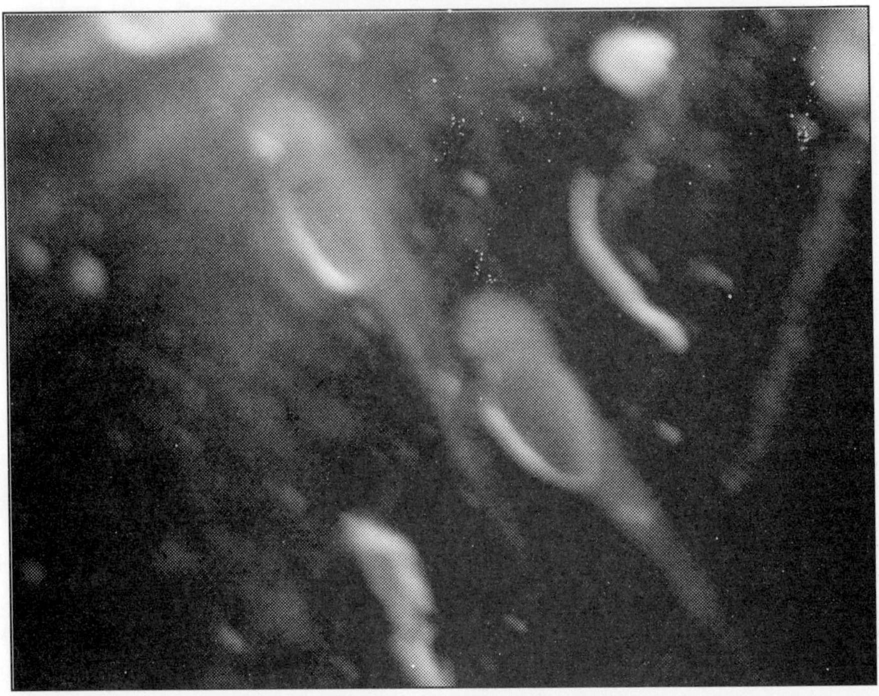

Plate 184: Two-Day Old Pacu Spawn

Plate 185: Pacu Fingerling

Plate 186: Pacu Fingerling Used as Ornamental Fish

time of initial injection. After 6 hours, final dose to be injected @ 12 mg per Kg body weight in female and 2 mg per Kg body weight incase of male and kept in a breeding pool for courtship. After 6-7 hours of final dose administration striping may be done for fertilization. Fertilized, water harden eggs are comparatively small in size. Eggs are slightly sticky in nature. To remove the stickiness we are using "multani mati" (available in grocery). Fertized egges then can be transferred to the hatching pool with gentle water circulation.

In order to achieve higher table fish production, apart from traditional pisciculture with six species, some more new species might be introduced with higher growth rate, better taste and standard food value which not only supplement the fish production quantitatively but enhance the earnings from a certain unit area of water body without hampering the ecological balance. Commercialization of any new economically viable fish for aquaculture needs some study, time and experiences. General study among the farmers in the locality of Naihati, West Bengal revealed that this sp. not only grows well up to 1.5 kg in one year with a survival rate approximate 90 per cent but also prove itself as a compatible species in a composite fish culture pond. In case culture op Pacu, along with *Cyprinus* gives better result in economic point of view. As it already gain a good market, widespread farming of Pacu is essential as a potential candidate species in aqua farming. So Pacu farming may be recommended as a new member for commercial aquaculture practice and Govt. may come forward to promote their culture.

(*iii*) Catfish Culture

Air-breathing catfishes have higher market value that the carps and enjoy higher consumer preference in many parts of the country. They are thus excellent candidate species for culture as well as for increase profit margin. The most commonly cultured catfishes in the country include magur (*Clarius batrachus*) and singhi (*Heteropneustes jossiUs*). These species are cultivated in many regions, these are preferably used in the culture ponds of Indian States, mainly as a candidate species in makhana cum fish culture systems and in swamps due to their adaptability to adverse ecological conditions. Commercial culture of air breathing fishes has great potential considering the availability of huge resources in the form of swamps and derelict waters. While polyculture of these species with Indian major carps is the usual practice, their monoculture has also been found to be very promising. With more or less similar pond management measures as that of carp culture, stocking with 20,000-50,000 fingerlings/ha and production levels of 3-5 ton are achieved in grow-out culture of magur, which attains 100-200 g each in 6-8 months. Apart from these two catfishes, culture of few catfishes like *Ompok pabda* and *P. pangasius* in the freshwater ponds of the in our state is having high potential. Though breeding technologies for these species have been standardized in North Bengal as well as in Naihati area, their mass seed production on a commercial basis is still awaiting which require some Government support in respect of funding and marketing.

(*iv*) Culture of Medium and Minor Carps as a Diversification

Promising medium carp species *viz. Labeo calbasu, L. bata, L. gonius, Puntius sarana* may be included with the carp culture in the main stream of farming in the state.

Puntius gonius showed good growth performance along with carp culture and auto-breeding habit may be included for species diversification in the freshwater ponds.

Minor carp species enter into the pond system through the run off, breed in captivity. Their adaptability to wider water characteristics help them to establish a good population in it. Some of the potential species available include *Puntius sp., Oxygaster sp., Gudusia chapra, Amblypharyngodon mola.* Polyculture of these minor carps along with carps may be successfully produced in all carp grow out ponds.

(v) Diversification of Aquaculture with Ornamental Fish

Culture of ornamental fish is one such diversification that involves the culture and propagation of fish and other aquatic organisms for its economic and aesthetic values. Considering the increased popularity of ornamental fish at household level, captive seed production and rearing technology of both domesticated exotic species and indigenous species have been standardized. Thye main advantages of ornamental fish culture are as follows:

- ☆ It is possible to do Ornamental fish culture in a much smaller space than normal food fish culture.
- ☆ It is possible to do Ornamental fish culture by investing much smaller capital than that of normal food fish culture.
- ☆ It is possible to do Ornamental fish culture using lesser manpower than normally required for food fish culture.
- ☆ The flexibility in the business of ornamental fish is enormous when compared with normal pisciculture. It may range from few thousand to multi million. In other word while discussing its scope one may that starting from ground only sky is the limit.

(a) Ornamental Fishery Scenario

- ☆ Ornamental fishery has a very bright prospect not only in the domestic market but has tremendous export potential as well.
- ☆ West Bengal is pioneer in ornamental fish culture
- ☆ West Bengal is no. 1 in Export
- ☆ West Bengal is no. 1 in Internal market
- ☆ 70 per cent share in internal market distribution
- ☆ Highest in human resource exploitation. 20 to 30 thousand women fishers are involved
- ☆ Highest in women self help group involvement. 132 such women-groups are running successfully in this state–an unique involvement in our country. Better recovery from these women SHGs than male fishermen groups.
- ☆ There are 400 varieties of ornamental fish of significant importance in comparison to only 20 no highly important commercial varieties of food fish.
- ☆ To explore the possibility of ornamental fish culture, Fisheries department has introduced this programme under F.F.D.A and NCDC programmes the

Plates 187–188: Ornamental Fish Farm

Plate 187

Plate 188

districts of North 24 Parganas and South 24 Parganas have been covered. Upto 2007-08, 852 nos. of ornamental fishery units have been developed.

(*b*) Basic Requirements for Ornamental Fish Culture

- ☆ Space/shade for setting the culture site
- ☆ Cement or plastic cistern/glass aquarium/majla/small pond.
- ☆ Availability of water and its distribution system.
- ☆ Water exchange facility on regular basis.
- ☆ Arrangement for aeration supply, lighting and electrification with standby facility.
- ☆ Fish seed with good Ornamental market value.

- ☆ Feed and feeding facility.
- ☆ Manpower for daily monitoring and maintenance of water quality and fish health.
- ☆ Miscellaneous article like bucket, mug, hand net, testing kits, etc.
- ☆ Technical knowledge.
- ☆ Prophylactics and measures for controlling disease problems.
- ☆ Packaging and selling arrangement.

(c) Culture of Green Shrimp, Black Shrimp

Occasional survey in the villages of North Bengal reveals a kaleidoscope of prawn species (*Macrobrachium scabriculum*, Green Shrimps (*Caridina* sp.), Black shrimp with brilliant color and patterns, which could be a demand in the ornamental fish market. Actually they are in great demand and certain collectors exploiting the ignorance and poverty of the local people and tribals collect them and minting dollars by exporting them to the international ornamental fish markets. Beside their ornamental value these species of prawns are edible too. They have considerable market demand as edible species. The breeding procedure of this said three species have been standardized by our Research Station at Kulia, Kalyani and now the seeds may be readily available for their culture in cages with an aim to explore their ornamental value as well as for food fish in the market.

10.3. Scientific Technology to be Followed in Intensive Fish Culture Method

The positive transformation in the business environment brought about by the economic reforms, programme undertaken by the state, the resurgence of the industry in the district, combined with the decisive advantage of rapidly growing, untapped market in this region availability of surplus power, low cost manpower developed agricultural economy for different fish crops throw up vast investment opportunities in the districts of North Bengal that can not be ignored. Investing in this region today makes sound business sense.

This has already generated a lot of enthusiasm and a number of investment proposals as are being received. At this juncture a better coordination, free flow of information and transparency in working of different organization and agencies would go a long way to set up fishing industry in this region. In respect of total industrialization, the concept of district level committees at the state level would perhaps help to achieve the desired degree of coordination and cooperation between different agencies facilitating new industrial ventures. Under these circumstances the entire North Bengal may become a prime industrial fishing region of the state.

The brief scenario of traditional aquaculture practices in the state is as follows:

- ☆ Indiscriminate stocking with seed or desirable and undesirable varieties of fish especially IMC
- ☆ Stocking of seeds in unprepared pond

- ☆ Stocking of seeds in ponds having predatory and weed fishes and aquatic weeds
- ☆ Lack of post stocking care
- ☆ Little or no fertilization or supplementary feeding
- ☆ Production absolutely depends on the inherent fertility of ponds *i.e.* available natural food

Aquaculture intensification is now required for enhancing production. In carp culture alone, various input based culture systems have been developed, namely traditional, extensive, semi-intensive and intensive, suiting to the region, resource characteristics, investment capacity and socioeconomic conditions of the farmers. Besides, the practice also differs based on management strategy such as single stocking-single harvest, Single stocking-multiple harvest, multiple stocking-multiple harvest or continuous culture, etc. Culture of Indian major carps only or polyculture along with exotic carps at low to moderate stocking density has been yielding a production of 4-10 tonnes/ha/yr. The technology of intensive carp culture involving stocking with advanced fingerlings of 50 g at higher densities of 20,000-25,000 nos/ha and provision of balanced supplementary feed, fertilization, aeration, water exchange and fish-health management has demonstrated higher production levels of 10-15 tonnes/ha/yr.

Pre-stocking Management

Pond Requirement

- ☆ Depth, Size and Shape
- ☆ Rectangular ponds are preferable for better oxidation of water, movement of fish.
- ☆ The size of the pond varies from 0.1-2 ha; depth varying from 2-3 meter
- ☆ Perennial ponds are more suitable than seasonal pond.
- ☆ Pond should receive a fair amount of sunlight.

Soil Condition

- ☆ The soil must be loamy or clay loamy.
- ☆ Bottom free from excessive organic matter.
- ☆ pH should be between 6.5 – 7.5.
- ☆ Available Nitrogen – 30 – 50 mg./100 gm.
- ☆ Available Phosphorous (P_2O_5)–6.0-16.0 mg./100gm.
- ☆ Organic carbon 1.0 – 2.0

Water Condition

- ☆ Turbidity bellow 20 ppm.
- ☆ pH – 7.0-8.0
- ☆ Total alkalinity –75 – 150 ppm.
- ☆ Nitrate (NO_3) and Phosphate (PO_4)–.02 -.05 ppm.
- ☆ Dissolved Oxygen – 5.0-10.0 ppm.

Corrective Measure

1. *Correction of acid soil*: By liming. Total quantity of lime may be divided into 3-4 installment and applied at 2-3 days interval.
2. *Correction of Sodic soil*: Highly alkaline soil may be corrected by applying cow dung @ high dose *i.e.* 20 –30 ton/ha and Gypsum ($CaSO_4$) @ 5-6 ton/ ha
3. *Correction of sandy bottom*: The fertility status and water retention of ponds with sandy bottom are poor. Application of organic manures *viz.* cow dung biogas slurry @ 20 –30 ton/ha help to improve the condition.
4. *Correction of heavy load of organic matter on the bottom*: Here corrective measures are by removal excess mud and liming @ 1 ton/ha and expose the pond bottom to the sun for about a fortnight.

Stocking Management

Rainbow trout, *Salmo gairdn*eri are natives of the Sacramento river region on the West Coast of United States of America, but have been successfully introduced in our waters. Similarly Brown trout, *Salmo trutta fario* are natives of European water have also been introduced in our waters. Rainbow trout is perhaps the best suited salmonoid for commercial production in the upland of Darjeeling as it can be easily domesticated. It accepts artificial feed and can withstand temperature between 20°C and 22°C. It can have a transient survival at a temperature of 25°C. It is disease resistant and is having faster growth rate over the other members of the family (Sen and Biswas, 2006).

Optimal Climatic Conditions

- Temperature 16-28°C
- Relative humidity 70-80 per cent

1. Heat tolerance of duck is good. They endure even 40°C peal temperature off enough shading and open water for bathing is available. The shedding shed is to be large enough for all the ducks for comfortable resting.
2. The heat tolerance of the laying duck is worse compared to the rest of the stock due to the lower feed consumption entailing the decrease of egg production. With young duck heat peaks do not cause any problem because of their excellent tolerance.
3. Duck is sensitive to great heat fluctuation only in the first two weeks, especially to very low temperature in the first 10 days of their age.
4. With rearing the duck in fish ponds, a lot of factors influence the number of duck harvested from 1 sq.m. annually; speed of water flow, water dept. biological status of the water etc. on the basis of practical experiences under optimal conditions 1 –3 ducks may be reared per sq.m. of water surface annually.

Selection of Fish Species is Important as it Decides Ultimate Fish Production

- Quantitative production is highest with herbivorous, omnivorous, plankton eating and detritus feeding fishes.

Plates 189–192: Fish Species Selection for Stocking

Plate 189: Paku

Plate 190: Bhetki

Plate 191: Magur

Plate 192: Pungus

- ☆ Introduction of fish species like Pungus/Paku having higher growth rate and market demand replacing Mrigel and ordinary Rohu respectively yield higher production within stipulated period.
- ☆ Stocking Density may be 30,000 per ha with a size of 10cm. (Phase-wise multiple or single stocking as per culture strategy)
- ☆ Ratio of Surface, Column and bottom feeding fish is 3:4:3
- ☆ Other species that can be stocked in small numbers in composite fish culture, are *Ompok bimaculatus, Mystus seenghala, Notopterus chitala,* the grey mullet (*Mugil cephalus*) and *Channa maruli.*
- ☆ *Mystus aor* and *Clarias batrachus* have also promising result in composite fish culture.
- ☆ For improvement of stock in culture fishes of Darjeeling hills, minor carps and English carp may be introduced in addition to the existing ones. This will enhance production per unit area.

Fish Seed Farm

Quality fish seed are the essential pre-requisites for successful fish farming. In order to get quality Fish seed, a fish breeding farm equipped with eco-hatchery with the provision to rear fish seed up to 4-5 cm. have to be established in the lower reaches where the range of temperature is high (as the metabolic intensity of fish is directly proportional to temperature) Areas at lower reaches near the river will be suitable provided it satisfies other requirements.

Transit Farm

Transit rearing farms in the Block Headquarters are required to facilitate distribution of fish seed to the fanners staying in the remote rural areas. These farms will be equipped with Oxygen cylinders, polythene bags and other requisites for packing of fish seed for proper transport.

Stocking Rate for Intensive Aquaculture Practices

Carp culture Rohu–95 per cent + Catla 5 per cent

Carp culture with complementary species -Rohu 90 per cent + Catla 4 per cent + Murrel 1 per cent (or)

Rohu 90 per cent + Catla 5 per cent + Murrel/Magur/singhi 5 per cent

Catfish (Pangas) culture -Pungus mono culture (or) Pungas 90 per cent + Rohu 10 per cent combination

Feeds to be Used for Intensive Aquaculture

Formulation of well-balanced diet and their adequate feeding are two of the most important requirements for successful fish farming. A constant supply of energy is necessary for all animals to sustain life. Energy for all animal systems comes from the feed eaten or in times of feed deprivation, from body stores. Until recently the emphasis was on natural food stuff and still largely depend upon natural food with some supplementation with the waste product of other form of Agriculture or industry *e.g.* slaughter house waste, grain waste, silk worm pupae and night soil. At low density *i.e.* in extensive system of culture these diets are adequate as most of the other nutrients of fish are satisfied from natural sources.

However, at high density like intensive culture in running water Jhora fisheries of Darjeeling hills, fish are dependent on natural food and some quantity of locally available supplementary feed. Thus in intensive fish culture inadequacy of supplementary feed lead to poor growth, nutritional disease and in poor fish condition, increased susceptibility to parasite and bacterial infection. Optimum dietary protein levels for common carp are 30-38 per cent whereas that of Rainbow trout is 35-45 per cent. There are some 40 essential nutrients that must be included in the diet of fish. The fishery section under Darjeeling Gorkha Hill Council should negotiate with the reputed company to arrange for supply of pelletised balanced food to the farmers undertaking intensive carp culture or trout farming (Sen and Biswas, 2006). The feed ingredients for intensive farming may be as follows.

Culture	*Feed Ingredients*	*Ratio*
Carp seed	Rice bran + GNOC/Soycake	2:1
Carp growout	DORB+GNOC/Cotton cake/Sunflower	80:20
	Boiled RB+Broken rich +Maize/Jowar	15:65:20
Catfish (Pungus)	Boiled RB+Broken rich +Maize/Jowar	15:65:20
	Palm oil mill effluent +Bone meal	
Scampi and Shrimp	Formulated branded feeds	

Plate 193: Feeds getting Boiled at Farm Side in Andhra Pradesh

Input Applications in Aquaculture

Fish	Input	Dosage (ton per ha)
Carps	Farmyard manure	20-25 (or)
	Poultry manure	12-15 (or)
	Biogas slurry	30-40 (or)
	Vermicompost	8-12
	Bio-fertilizers	1-2
	DAP with Cow dung	100 kg
Prawn	Farmyard manure 1.0	

Post-stocking Operations

(a) Regular Monitoring of Physicochemical Parameters of Water

Fertilization and manuring accordingly

Supplementary feeding

- ☆ Artificial feeding enhances fish production and attributes 50 to 80 per cent of total fish production in ponds to artificial feeding.
- ☆ Extent of intensive feeding is an economic question which depends on the cost of the foods and their conversion rate termed also "food quotient".
- ☆ Food quotient varies considerably with variations in temperature, oxygen content of the water, size of the fish, feeding habits and general condition.

(b) Water Management

- ☆ Regular monitoring of Temp., DO, Alkality, pH, H_2S, Ammonia, NO_2-N.
- ☆ Maintenance of steady Zooplankton population 2-3 ml per 50 lt. of water or 30 cm Sachi disk transparency.
- ☆ Water depth in average is 2 – 3 mt.
- ☆ Water should be drained out during monsoon to maintain the desirable depth.
- ☆ Change of pond water or refilling has been done to increase productivity.
- ☆ Reconditioning of water by biological filter and aeration are done in the summer countries.

(c) Harvesting

- ☆ Multiple harvesting to be done from 3 month onwards of stocking
- ☆ Fishes over 500 gm. are regarded as marketable fish, can be harvested.
- ☆ Generally the fishes individually attain above 1 kg. in 12 months when they are harvested finally.

10.4. System Diversification in Fish Culture

(*i*) Recirculatory System

Recirculatory system of aquaculture is a suitable method of culture in regions of water scarcity and in areas where disposal of waste is a problem. Used water from ponds is recycled after passing it through a waste treatment system. Such a system has promising future for the developement of an industrial "aquaculture system in metropoliton cities.

(*ii*) Flow-through System

This particular system may be adopted for intensive aquaculture system where flow through can be achieved by gravity, particularly bellow the dams and other river valley projects. Water resources from streams, springs can be utilized for such type of system especially for carnivorous fish culture *viz.* catfish living in higher level of food chain.

(*iii*) Organic Farming

Production of organic fish is a new avenue in carp farming in recent years. The main problem in this aspect is availability of manure. This problem can be solved if the farmers of the state adopt integrated fish farming as per the availability of their land resource.

(*iv*) Sewage Fed Fish Farming

This system has already been well adopted in the East Kolkata wetlands in large scale and few districts in small scale in the state. The productivity is very lucrative. However, there is a huge scope to adopt this farming in the most of the Municipal area of the state for recycling the sewage water and conserving energy. Multiple stocking and multiple harvesting are the main approaches in this system.

(*v*) Cage Culture

Species which have successfully been reared and grown in cages at Pailan fish farm under the ARHMC are Bhetki (*Lates calcarifer*), Pacu (*Piranha sp.*), Pangus (*Pangusius sp.*), Magur (*Clarius batrachus*), Golden Mahaseer (*Tor puntitora*), Sarputi (*Puntius sarana*), Monosex Tilapia (*Tilapia sp.*), Post larvae of Freshwater giant prawn (*Macrobrachium rosenbrgii*), some ornamental fishes like Koi carp (*Cyprinus sp.*), and Gold fish.

Cage culture of fish is a method of raising fish in containers enclosed on all sides and bottom by materials that hold the fish inside while permitting water exchange and waste removal into the surrounding water. Cages range in size from 1-m^3 to several hundred m^3 and can be any shape. The minimum recommended stocking density for common carp, tilapia, Paku and catfish is 80 fish/m^3. The smallest recommended fingerling size for stocking is 15 g. High prized fish like Sea bas (Bhetki) that cannot be cultured with other fish due to its predatory habit can be cultured in cage in same pond porofitably. Adequate supplies of good quality feed must be given to the fishes. Non-filter feeding fish confined in cages have limited access to natural pond foods and need a nutritionally complete diet. When feed is provided, water

Plates 194–195: Cage Culture Demonstration at ARHMC, Pailan

Plate 194

Plate 195

exchange is needed to bring oxygen into the cage and to remove waste products generated by the fish as a result of feeding. The weight of fish produced in cages depends on many factors including the fish species, stocking density, fish size at stocking, culture period, cage size, water quality, and feeds used. Different types of diversified aquaculture is described bellow.

(a) Cage Culture of Seabass–Bhetki

- ☆ Specifications of the Cage-Floating, made up of synthetic net with mesh size 1 mm., sealed P.V.C. pipes are used for buoyancy of the cage.Size: 4.5 x 2.5 x 1 mt.
- ☆ Culture period: 90 days.
- ☆ Seed size: 5-6 cm. Weight: 6-7 gm
- ☆ Stocking Density: 30 pcs/mt^3
- ☆ Trash fishes, small shrimp may be fed @ 20 per cent of the body weight at initial stage and gradually reduced to 10 per cent of body weight daily in the morning. In the afternoon inert feed composed of boiled chicken flesh waste, and formatted feed may be fed @ 5 per cent of body alternately.
- ☆ Regular weekly health checking.
- ☆ Cages to be cleaned manually once in a week.

(b) Cage Culture of *Pangasius*

- ☆ *Pangasius sutchi* can be produced in cages.
- ☆ The volume of wooden cage is 15 m^3 to 20 m^3 with width–2 m, length–5 m, depth–1.5 m, suspended in running water.
- ☆ Grow-out cages are stocked with about 50 fingerlings, 8 to 10 cm long, per cubic meter of cage.
- ☆ Mortality is relatively low.
- ☆ Marketable size is about 1 kg, and it takes approximately 15 months.

(c) Cage Culture of Magur

- ☆ *Clarius batrachus* can be reared in cages for 6 months.
- ☆ Stocking density may be 8 pieces per cft with average size of 2 cm.
- ☆ The fishes should be fed twice daily with mixture of broken rice, rice bran, broken pea and corn in equal ratios @ 7-8 per cent of body weight. Supplemented with groundnut oil cake once daily @2 per cent of body weight.
- ☆ Physico-chemical parameters within cage and outside cage *i.e.* pond water to be tested regularly 1-2 days in a week.
- ☆ Samplings may bemade twice in a month.
- ☆ Regular weekly health checking should be done.
- ☆ Cages to be cleaned manually once in a week.
- ☆ Culture period–180days and expected average body weight–100 gms.

(d) Cage Culture of Paku

- ☆ *Piaractus brachypomus* can be reared in cages for 6 months.
- ☆ Stocking density can be 4 pieces per c ft. with average size of 2.5cm.
- ☆ The fishes may be fed twice daily with boiled mixture of broken rice,ricebran,broken pea and corn in equal ratios @ 7-8 per cent of body weight. Supplemented with ground nut oil cake once daily @2 per cent of body weight.
- ☆ Water quality parameter within cage and outside cage to be tested regularly.
- ☆ Samplings may be done twice in a month.
- ☆ Cages to be cleaned manually once in a week.
- ☆ Culture period–180days and expected average body weight–200 gms.

(*e*) Cage Culture of Sar Punti

- ☆ *Puntius sarana* may be reared in cages for 200days.
- ☆ Stocking density may be 12 pieces per cft with average size of 4.5 cm.
- ☆ The fishes may be fed twice daily with boiled mixture of broken rice,ricebran,broken pea and corn in equal ratios @ 5 per cent of body weight. Supplemented with ground nut oil cake once daily @2 per cent of body weight.
- ☆ Samplings may be made twice in a month.
- ☆ Culture period–200days and expected average body weight–45 gms

(*f*) Cage Culture of PL of *Macrobrachium rosenbergii*

- ☆ Stocking–15 PL per cft
- ☆ Twice Daily @ 10 per cent body weight.
- ☆ Aeration may be done twice for 2-3 hours in the late night and evening with paddle aerator.
- ☆ Weekly sampling and health checking.
- ☆ Change of cage net-every 15 days
- ☆ After 60 days rearing Prawn Seed grows 6 cm average size with survival rate 80 per cent.

10.5. Integrated Fish Farming

The main objectives of the integrated fish farming are as follows:

- ☆ Establishment of man-made ecosystem without any wastes.
- ☆ Promote organic farming.
- ☆ Increasing the food supply for the mankind.
- ☆ Reducing the cost, raising the output and increasing economic benefits.
- ☆ Promote finance and support producer owned and controlled organizations.

☆ Facilitate Transformation of traditional farming system into a commercial and technology based sustainable enterprise to maximize farmer's farm income.

☆ Facilitate the growth of agribusiness in the state for Farmers, Social, Corporate and Environmental Good.

☆ More employment generation.

☆ Facilitate the generation of increase on-farm and off-farm employment.

I. Different Types of Integrated Farming in West Bengal

Integrated fish farming is a diversified and co-coordinated way of producing agriculture crops and livestock products on fish farms with fish as the main product. In West Bengal the integrated fish cultured model may be introduced as per given below:

	*Area**
Aquaculture–Livestock integration	
Pig cum fish farming	HA, NZ, CZ, WZ, EZ
Poultry cum fish farming	NZ, CZ, WZ, EZ
Duck cum fish farming	NZ, CZ, WZ, EZ
Cattle cum fish farming	NZ, CZ, WZ, EZ
Goat/sheep cum fish farming	NZ, CZ, WZ, EZ
Rabbit/Guinea pig cum fish farming	NZ, CZ, WZ, EZ
Termite cum poultry cum fish farming	NZ, CZ, WZ, EZ
Fodder cum cattle cum fish farming	NZ, CZ, WZ, EZ
Aquaculture-Agriculture Integration	
Rice cum fish farming	WZ, EZ
Rice cum Azolla cum fish farming	WZ, EZ
Aquatic plant cum fish farming	NZ, CZ, WZ, EZ
Mulberry/silkworm cum fish farming	CZ, WZ,
Horticulture cum fish farming	NZ, CZ, WZ, EZ
Fodder cum fish farming	NZ, CZ, WZ, EZ
Mushroom cum fish farming	HA,NZ, CZ, WZ, EZ
Orchids cum fish farming	HA
Waste water fish farming	NZ, CZ, WZ, EZ
Bio gas slurry cum fish farming	NZ, CZ, WZ, EZ
Vermi-compost cum fish farming	NZ, CZ, WZ, EZ

* HA: Hill Area [Darjeeling, Jalpaiguri (some part)]; NZ: Northern Zone [Cooch Behar, Jalpaiguri,Darjeeling (except hill areas), Uttar Dinajpore and Daskhin Dinajpore]; CZ: Central Zone [Maldah, Birbhum, Murshidabad, Nadia]; WZ: Western Zone [Purba Midnapore, Paschim Midnapore, Purulia, Bankura]; EZ: Eastern Zone [Burdwan, Howrah, South 24 Parganas, North 24 Parganas, Hooghly].

II. Brief Description

The brief description of first four models is depicted bellow.

(a) Pig-cum-Fish Farming

About 70 percent of pig manure appears to be readily digestible by fish. Food, while passing through the alimentary canal of the pig, gets mixed with enzymes that continue to digest voided manure. Therefore, pig manure with its higher percentage of digestible portion and enhanced enzymatic action, could serve as a beneficial food source for fish.

- ☆ The raising of pigs can fruit-fully be combined with fish culture by constructing animal housing units on the pond embankment or over the pond in such a way that the wastes are directly drained into the pond.
- ☆ The number of pigs required will depend upon the pond area. The excreta of three pigs are sufficient to fertilize a pond of 1 000 m^2.
- ☆ So, thirty pigs (30) may be raised on a pond of 1 ha. As pigs attain slaughter size within 5-6 months and fish raising of Indian exotic carp is done for 10-12 months, two lots of pigs can be raised along with one lot of fish.
- ☆ The pig food are often leftover and residues from the kitchen, aquatic plants and agricultural products and their wastes. The pig excreta is in turn used as organic manure in fish ponds.
- ☆ The C:N ratio of pig excreta is about 14:1 and the conversion of fresh pig manure to fish is about 25 kg/1 kg of fish.

Plate 196: Pigsty Constructed Over an Embankment

Plate 197: Pig Waste Directly Falling on the Pond Water

Plate 198: White York Shire Variety of Pig

Plate 199: Deshi Variety of Pig

- ☆ Due to abundance of natural food in the fish-pig pond, the fish attains marketable size within a few months. Partial harvesting, therefore, should be done three times, depending upon the growth of fish. Final harvesting may be done after 10-12 months.

(*b*) Poultry-cum-Fish Farming

Chicken raising for meat (broilers) or eggs (layers) can be integrated with fish culture, which has certain advantages:

- ☆ The poultry droppings act as excellent pond fertilizer and raise the biological productivity of the pond and consequently increases fish production.
- ☆ Some of the fishes feed directly on the poultry droppings.
- ☆ No supplementary feed is required for the fish culture, which normally accounts for 60 percent of the total input cost in conventional fish culture.

Plate 200: Deshi Variety of Cock

Plate 201: White Leghorn Variety of Chicken

- ☆ It maximizes the use of space.
- ☆ Saves labour in transporting manure to the ponds and the poultry house is more hygienic.
- ☆ No significant differences have been observed on the chickens' growth or egg laying when they are raised over the ponds or on land. In case of the former, the pond embankment could still be utilized for raising vegetables.

(c) Chicken Raising

- ☆ Broilers: 30-50 broilers could be raised on a 1 000 m^2 pond
- ☆ The chicken house can be constructed over the pond at least 0.5 m above maximum pond water level, or on the embankment. Each bird requires 1.5 ft^2 space (50 birds require 75 ft^2 space).
- ☆ The house can be made of bamboo or any other locally available cheap materials.
- ☆ Roof can be covered with hay or similar material. Enough cross ventilation should be maintained to keep cool during hot days. Floors are to be constructed.
- ☆ Any fast-growing chicken, broilers, can be raised.
- ☆ Feed with starter mash for 1-4 weeks and with finisher mash for 5-8 weeks, given as much as they can consume. Water should be provided at all times.

(d) Duck-cum-Fish Farming

Raising ducks over fishponds fits very well with the fish polyculture system.

- ☆ Ducks fertilize the pond by their droppings when given free range over the pond surface.
- ☆ Ducks have been termed as manuring machines for their efficient and labour-saving method of pond manuring, which accounts for 60 percent of the total cost in conventional fish culture.
- ☆ Ducks keep water plants in check.
- ☆ Ducks loosen the pond bottom with their dabbling and help in release of nutrients from the soil, which increase pond productivity.

Plates 202–203: Duck as Biological Aerator in Fish Pond

Plate 202

Plate 203

- ☆ Ducks are the volunteer aerators. The swimming, playing and chasing of ducks in a pond more or less play a role in aeration. They have been called 'biological aerators'.
- ☆ About 300 ducks are sufficient to fertilize a pond of 1 ha and this number only needs a (house) floor area of 130 to 140 m^2.

The integration of fish-cum-duck farming practice has been developing very rapidly in recent years especially in water abundant areas. Mirik Lake is most suitable for such a large-scale integrated farming. The ducks will be let loose to allow free ranching in the in the lake during the day but will be kept in sheds/pens at night at 'Homidara' (island within the lake). This method is advantageous to this large water body for feeds. This method can be considered as an integrated management model for large water body development.

Table 115: Demand and Tolerance of Hungarian Duck in Various Life Periods

Age	*Temperature*
0 to 3 days	30–35ºC
4 to 7 days	23–29°C
8 to 14 days	15–22°C
From 15 days (max.41-42°C)	12–38°C

The integrated management of fish-cum-duck has been found to be one of the economically efficient farming practices from the view point of input-output relation–ship, integrated fish farming with ducks is considered to be the best model of integration of fish and livestock because of 'symbiotic' relationship of ducks.

(*e*) Cattle-cum-Fish Farming

Cow or buffalo are herbivorous ruminant animals, and can better utilize green forage. Milk cow or buffalo as ruminant animals, can well utilize them. That's why milk cow or buffalo are one of animals raised in "fish-livestock" integration. The milk output of the cow or buffalo is high, economic benefit great, and excreta considerable. The main approach of raising milk cow or buffalo is grazing.

- ☆ Both milk and fish harvests.
- ☆ Each cow can provide a total of 210.8 kg of protein that is 3700,000 kcal in calorie value annually.

Plates 204–205: Cattle Farming

Plate 204

Plate 205

- ☆ Fish-cum-cow integration can increase protein by 36 per cent and calorie value by 10.6 per cent above the unitary cow farming.
- ☆ Cow dung are readily available for fish which is an excellent manure for fish ponds.

(*f*) Horticulture

- ☆ Fruit plants like Papaya, Banana, Coconut, Vegetable plants like Brinjal, Cabbage, Cauliflower, Tomato, Cucumber, Pumpkin, Ashgourd, Bottle gourd, Radish, Beans, Spinach, Cowpea, Okra, Colocasia and other leafy vegetables can be grown on the pond dikes.

Plate 206: Horticulture in Fish Pond Dyke

- ☆ The success of the system depends on the selection of plants.
- ☆ They should be of dwarf type, less shady, evergreen, seasonal and highly remunerative.
- ☆ Dwarf variety of fruit plants like mango, banana, papaya, lime is suitable as they do not obstruct sunlight to water bodies.
- ☆ Pineapple, ginger, turmeric, chilly are grown as intercrops.

10.6. Measures to be Adopted for Sustainable Aquaculture Especially in Hill Region

The potentialities in sector of fisheries in North Bengal especially the coldwater fisheries are required to be explored in the interest of sustainable development through effect under the supervision of Dept. of Fisheries, Govt. of W.B.

1. To generate skilled manpower in the sector of coldwater fisheries, to establish training centers, and carry out culture projects at the locally available coldwater fishes like *Acroscheilus dyochillus*, Indian snow trout (*Oreinus noleswerthini*) will be adopted at same site.
2. To conduct survey work to reveal the abundance of variety of fishes in the different water systems of these 2 districts.
3. To introduce model scheme with aims and objects to help the poor fish farmers financially and to train them to culture fish and to assist them with bank loan.
4. For self-sufficiency of fish seed, construction of a fish seed firm and eco-hatchery for production of quality fish seeds.
5. In a novel effort under the supervision of Dept. of Fisheries, West Bengal, the potentialities in the sector of cold-water fisheries are going to be explored in the interest of sustainable development.
6. The main theme of conservatory strategy is to create awareness in the people about the need of conservation. The imminent danger of extinction of natural resources needs to be explained to the people so that they themselves extend their helping hand for conserving their natural resources instead of causing damage.

With a view to utilizing this objective a state level aquarium has been installed, which will facilitate the viewers to acquaint themselves with hatchery resources and stimulate a feeling about the need of their conservation. Lane of the measures which can be taken to prevent pesticide pollution is:

1. Judicious use of pesticides.
2. Continuous monitoring in the environment.
3. Pesticide with less lethal effect should be prepared.
4. Pesticides with bio or phase degradable residue should be preferred.
5. Topographical barrier on to is made to stop the spillage from agro-field into the ponds.
6. Biological control measures have to be taken up whenever possible.

10.7. Guiding Principles for Sustainable Development

Conservation of natural resources; Integrated development of natural as well as social resources; Insitu moisture conservation; Sustainable farming system; Adoption of ridge to valley approach; Due emphasis on production enhancement activities for land owners and livelihood support for landless families; Democratic decentralization in decision making; Transparency in transactions; Mobilization of community at the village level; Direct funding to the community; Emphasis on "Government" participation in "Community" plans; Contributory approach to empower the community; Building upon indigenous innovations; Initiatives and ideas; Equity for resource-poor families and empowerment of women; Moving away from subsidy oriented development to self-reliant development; Convergence of activities/schemes of government and non-governmental organizations etc.

District Level Nodal Agencies will prepare a list of all eligible villages in each Block which could be considered under the watershed programme on the basis of various parameters indicated above.

For a unit watershed of 500 ha, the budget provision of Rs.22.5 lakhs and Rs.30 lakhs has been made for areas having less than 8 per cent and more than 8 per cent slope respectively. A broad allocation of funds for major components is given in Table 116.

Table 116: Allocation of Funds for Major Components

Sl.No.	*Components*	*Allocation of Funds (%)*
A.	Management Component:	
1.	Administrative cost	10
2.	Community Organisation	7.5
3.	Training Programme	5.0
	Sub Total	22.5
B.	Development component:	
1.	Natural Resource Management	50
2.	Farm Production system for land owning families	20
3.	Livelihood support system for land-less families	7.5
	Sub Total	77.5
	Total	100

The ultimate objective of the National Watershed Development Project for Rainfed Areas is to make the watershed slogan "Kheti ki mitti khet mein, khet ka Pani Khet mein; Gaon ka mitti gaon mein, gaon ka Pani Gaon mein" a reality in the rainfed areas of the country through people's movement. As the above project has made avenue for the watershed management, an effort need be taken to utilize the same in the hills areas for the benefit of the local people, agriculture, aquaculture and tourists.

10.8. Alternative Livelihood for Better Tomorrow

Capture fisheries being capital intensive and less productive, aquaculture is nowadays considered to be the appropriate proposition for production of fish. There are three main ways to increase production *viz.* Intensification, Integration and diversification of aquacuture practices.

1. At the outset, the production has to be increased in a sustainable manner to bridge the dietary gap in the state. The magnitude of fish production from aquaculture has to be intensified instead of traditional aquaculture in the state.
2. The flowing water may be used for pen culture.
3. Culture in cage with suitable/durable material may be developed for fattening of selective species.
4. In the surrounding sweet water pond, fishes like Magur (*Clarias batrachus*), Shinghi (*Heteropnaeustes fosilis*), Koi (*Anabus testudineus*), Shol (*Channa straitus*) may be cultivated by the fisher women otherwise professionally unengaged. More and more air breathing fish culture has to be promoted to combat the situation of water body shrinkage due to effect of climate change.
5. Paddy cum fish culture, lotus and trapa culture, algae culture along with fish in fishponds may be introduced for maximizing the income.
6. Integrated farming (Mixed Fish Farming -duckery -piggery -poultry -goatory) be intensified for the purpose of proper exploitation of food materials of all ecological niches in the aqua system by compatible group of fishes and integration of other input for maximum output.
7. To meet up the requirement of seed for fresh/brackish water aquaculture, multispecies and multipurpose hatchery-cum-nursery be developed.
8. Involvement of women in fishing industry by forming SHG be a task of the fishery personnel for exploring the vast market of world ornamental fish trade. Fisheries Department, GOWB has already made 132 Women SHG which is a record not only in our country, even in abroad.
9. Training for entrepreneurship should be a major priority. Women are already engaged in small trading enterprises. They should profit from inputs on risk taking, better business practices, and financial management.
10. Financial assistance in the form of Bank loan and Govt. subsidy for alternate income generation programme for seed catchers.
11. Govt. can purchase all the catches at fair price through seed collector's and can start model canal fishery culture with those seed. Huge loss of bio-diversity occurs during first seed collection, which may be prevented by utilizing tidal water fed follow canals for culturing unwanted fish seed, which are thrown away other-wise. Culturing of this fish seed is to be done through organized co-operative system.
12. Decasualization of fishery workers/fishermen during lean period may be thought under a special programme of constituting Fishery Labour Board in accordance with the prevailing provisions of the existing Regulation of Employment Rules. Decasualization model for Agri labourers may be taken up for fisheries sector too.

It is needless to mention that in inland fisheries, whether capture or culture, the yield gaps are significant and need to be filled in order to realize the full potentials.

Besides enhancing the fish production, production activities can be diversified through integration with agriculture and allied sectors. It will help in optimizing the water productivity using these farming systems and utilisation of derelict waters with a focus on food fish, high value species, ornamental species and also open the scope for establishing more aqua related ancillary industries in the state. The entire frame of activities of fisheries development has to be re-oriented with bulk emphasis on promotion of pisciculture by ensuring people's participation on a larger scale in the pisciculture activities.

Chapter 11

Research and Developmental Needs for Rebuilding the Natural Aquaspecies

11.1. Research-Extension Initiatives for North Bengal

What we Lack

- ☆ Almost non-existent multi-stakeholders' partnership in designing and implementing the rural mainstreaming research in a system perspective
- ☆ Absence of effective research into use mechanism for integrating different technology-led livelihood opportunities with the market and delivery systems
- ☆ Initiatives pertaining to validation of available research information and technology outputs under widely varying aqua-ecological and socio-economic regime
- ☆ Community education pertaining to compliance of quality parameters and certification procedures
- ☆ Extreme dearth of skilled rural human resource for effective knowledge networking
- ☆ Initiatives in partnership building with potential rural communities in production of quality seeds and other inputs

The following research and developmental problems are being suggested in addition to existing one in the context of Capture and Culture Fisheries.

11.2. Capture Fisheries

1. Scheme for conservation of Giant Himalayan Mahaseer in the hill streams of Darjeeling by establishment of hatchery at a suitable place. Kulkurni (1977) has succeeded in breeding *Tor khurdree* by stripping in Walwhan and Shirawata reservoirs of Lonavala, Bombay and rearing its eggs and hatchlings in running water stream. TATA Powers has also successfully established Mahaseer Hatchery. The procedure will be more or less similar and the mahaseer seedlings will be liberated back to the streams.
2. Study of carrying capacity of hill streams to evolve stocking policies to make sport fishery more attractive by exploring the possibilities of liberation (ranching) of trout in some of the streams by studying the physico-chemical and biological characteristics of hill streams.
3. Implementation of conservation measures to Asla, *Schizothorax progastus* and *Schizothorax plagiostous* and Gurdi, *Labeo dero* and *Labeo dyochillus* in the hill by way of setting up of hatchery and ranching of fingerlings and effective banning of catching of juvenile and brood fishes in the streams.

11.3. Culture Fisheries

1. Establishment of experimental Trout farm by starting semi-intensive culture by introducing disease resistant, available improved fast growing varieties of rainbow trouts *viz*, Donaldsons and Hampshire in a suitable site for upland water with hatchery facilities.
2. Arrangement for pelleted feed for intensive culture of carps (catli, common, grass carp, rohu etc) in particular in running water pools/ponds, fed by perennial jhora as well as for the trouts.
3. Setting up of fish seed farm and eco-hatchery in lower reaches for selected varieties of Indian Major Carps *and* exotic carps like common carps and Grass Carps with provision to rear it up to advanced fry stage (4-5 cm).
4. Setting up of ornamental fish seed farm for rearing of the advance fry to fingerlings stage in the eight Block Head Quarters for the benefit of seed supply to farmers. Towards this some Transit Seed Farms have been established in the Blocks and one eco-hatchery has been set up at Reyang by the State Fisheries Deptt. More such need to be set up to meet up the need.
5. Introduction of new species of common carp variety specularis *i.e.* mirror carp and English carp or golden carp, carrassius in addition to the existing species *i.e.* common carp variety communis and grass carp.
6. To enhance production/unit of Jhora Fishery by resorting to more stocking density.
7. Experiments on intensive culture practices of common carp and other fish species with energised pelleted feed and subsequent transfer of technology to the farmers.

8. Establishment of a duckery at Mirik Lake in 'Homidara' within the lake with Hungarian ducks.
9. To issue angling permits to the interested anglers against fees by fixing bag limits and size limits of fish in order to attract tourists and genuine anglers in the Mirik Lake and in the Rivers.

11.4. Research Priority

The following aspects may be considered for scope of research on priority basis in the uplands of Darjeeling District.

1. Study of carrying capacity of hill streams to evolve stocking policies.
2. Physicochemical and biological characteristic of three types of streams (Fast, medium and slow) in the upland of Darjeeling have to be studied.
3. Setting up of fish seed farm and eco-hatchery in lower reaches for IMC *and* exotic carps.
4. Arrangement for pelleted feed for intensive culture of carps.
5. Establishment of experimental Trout Farm by starting semi-intensive culture.
6. Experiments on intensive culture practices of common carp and other fish species.
7. Conservation measures in the hill by way of setting up of hatchery and ranching of fingerling.
8. Introduction of new species of common carp.
9. Scheme for conservation of Mahaseer by establishment of hatchery at a suitable place.
10. Setting up of ornamental fish seed farm.
11. Construction of awareness centre.
12. Blockwise resource study and production estimation in all districts of North Bengal.
13. Aquabiodiversity study in big waterbodies (Beels) of North Bengal.
14. Explore the scope for eco-tourism.
15. Scope for establishment of Aqua sanctuaries.
16. Scope for establishment of fish based industries and development of market linkages.

11.5. Outreach Challenges and Main Focus Required in North Bengal

- ☆ To device an umbrella approach for ensuring cross-sector partnership intervention combining resources.
- ☆ To develop and validate target groups specific technology modules designed in farming system approach across agro-climatic zones in cluster basis.

- ☆ To blend appropriate value addition technologies over time, space and form in compliance with critical certification standards and norms.
- ☆ To strengthen livelihood capacities of target groups, including women, for better empowerment.
- ☆ To facilitate the process of linking livelihoods with the market.
- ☆ To ensure quality service and expand the delivery outreach up to the neighbouring SAARC countries by introducing media mediated information communication system, using e-highway.

11.6. Approach Roadmap

- ☆ Cluster based cross-sectoral partnership with positioning of knowledge and dissemination partners and the process institutionalization.
- ☆ Collectivization of household capacities.
- ☆ Diversification of livelihood options through promoting integrated farming and fish based industries.
- ☆ Diversification of livelihood opportunities on the basis of strengths and opportunities of the clusters.
- ☆ Product development through .information facilitation and capacity building of SHGs.
- ☆ Addition of further values in products through registration and certification instruments.
- ☆ Facilitation for business development with appropriate linkages and institutions.
- ☆ Demand driven knowledge outsourcing.
- ☆ System analyses and process research.
- ☆ Knowledge networking at local level, regional level, extra regional level.
- ☆ Process registration and certification under MS ISO 9002.

Chapter 12

Strategies for Optimization and Prioritization of Species and Culture System

A State level seminar was held at Siliguri, Darjeeling, West Bengal on March 21-22, 2006 on the issue of Coldwater Fishery potential and prospects. The seminar was held at North Bengal University and jointly organized by the Department of Fisheries, Aquaculture, Aquatic Resources and Fishing harbors, Govt. of West Bengal and the Department of Zoology, North Bengal University, Siliguri, Darjeeling. Large number of participants, *viz.* faculty of the North Bengal University, faculty of Fishery Science of West Bengal, University of Animal and Fishery Sciences, Officers from the Department of Fisheries, Govt. of West Bengal and Sikkim; NGO's; fish farmers; entrepreneurs and fish traders; Director and scientists from National Research Centre on Coldwater Fisheries, Bhimtal and Central Inland Fisheries Research Institute, Barrackpore, Deputy General Managers from the NABARD, Divisional Forest Rangers, Darjeeling, Fishery Secretary and Fishery Officers from DGHC, Darjeeling and other stake holders of the area attended and contributed in this seminar (Plates 207–214).

The future strategies for increasing fish production from cold-water sector should be directed towards horizontal and vertical growth of the enterprise. Following components holds the key to success of these strategies subject to suitability of culture practices according to location and productivity of the coldwater.

Introduction of Good Management Practices (GMP) for Coldwater Fishes

Exclusive training is required on the field of breeding, nutrition, hatchery management and diseases of coldwater fishes. The technologies for the coldwater

Plate 207: Hon'ble MIC, Fisheries Meeting with Secretary, DGHC and Officials at Reyang

Plate 208: Sri S. Chakraborty, Special Secretary GOWB with Sri S. D. Sherpa, Secretary DGHC

Plate 209: Dr. M. Mukherjee, DF Delivering Lecture in CWF Seminar at Reyang

Plate 210: Dr. M. Mukherjee, DF with Trainees at Mirik

Plate 211: Dr. K.K. Vass, Director, CIFRI along with Coldwater Fisheries Scientists and Research Scholar

Plate 212: Sri R.F. Lepcha, JDF Delivering Lecture in CWF Training at Reyang

Plates 213–214: Participants of CWF Seminar at Reyang Fish Seed Farm

Plate 213

Plate 214

fisheries should be standardized and made available to the fish farmers. Study of carrying capacity of hill streams to evolve stocking policies and to make sport fishery more attractive by exploring the possibilities of liberation (ranching) of trout in some of the streams after evaluating the physico-chemical and biological characteristics of streams are required.

Intensification of Culture Practices

Culture practice may be intensified by following better management practices in ponds of North Bengal. The sub-temperature region in West Bengal is less productive as compared to the Indo-Gangetic plain region of the state. The productivity and production of food in high altitudes is comparatively less. The local people live on these natural resources being the main source of Iivelihood for them. The weaker sections of the society tend over exploit the natural resources, results in degradation. Therefore, the major objective of the development of aquatic resources in this region should be increasing productivity, production from aquatic resources to ensure nutritional food security at higher altitudes. The pristine sub-Himalayan ecosystems in this part are a source of global attraction and the development of fisheries should be based on eco-tourism. Therefore, it is strongly recommended that greater emphasis should be given to the development of food, sport and ornamental fishery. This will require full involvement of Ministers, Research Institutions, Development Departments and Voluntary Agencies/NGOs.

Diversification of Culture Practices

Diversification programme can be undertaken by adopting integrated culture systems. The coldwater fisheries resources comprising of varieties of fish spices including sport fishes in the hill streams in the district of Darjeeling could be developed thereby reviving the sport fishery to attract tourist as well as to propagate the population of some of the species marked endangered. Angling permits can be issued to the interested anglers against fees to attract tourists and genuine anglers in the Rivers. In order to ensure food and livelihood security of the hilly regions, high value species should be propagated both for on vertical and lateral expansion. Indian experience, over last 100 year has shown trout, snow trout, brown trout and other species have established at higher altitudes and in different ecosystems. Some of these trouts are exotic in origin but have been established in Indian waters. Hatchery system for these species has been developed in India. It is recommended that in West Bengal feasibility of rainbow trout needs to be evaluated. In Sikkim state-of-art trout, hatcheries should be established for propagation, culture and economic benefits. It is necessary to establish hatcheries and grow-out centers for stocking the natural ecosystems and glacial lakes in Himalayan region.

Introduction of Potential Fish Species

There is also tremendous scope for augmenting the fish production from the jhora fish pond by manipulating fish combinations of indigenous species like Mahseer (*Tor putitora*), Snow trouts (*Scizothorax sp*), Katli (*N. hexagonalipes*), Common carp (*Cyprinus carpio*), Grass carp (*Ctenopharyngodon Idella*).

Expansion of the Coldwater Resources

There is an urgent need to start up with the scientific approaches toward the assessment of present fishery status of rivers and jhoras. Importance should be given to the thorough study of biodiversity, water quality, and physico-chemical parameters of different streams to come up with suitable recommendations for the future ventures. It is necessary to conduct biodiversity surveys, identify and characterize the genetic resources by the use of modem techniques, the North Bengal University to initiate work on this line with active support from the State Fisheries Department. The resource assessment is undertaken through remote sensing with the collaboration of Department of Space and other organizations having expertise in the field.

Development of Resources for Captive Culture

Except a few individuals as a hobby, fish culture was not practised as a commercial proposition by any one in the hills till 1983. It was little realised that natural site for ponds are easily found on broken waste land, thus providing a means of utilising waste land, marshy land, derilict water meadows, gullies and ravines for conversion to small pools/ponds for fish culture. Depression on the ground, a succession of shallow little valleys or ravines, muddy water holes, all make excellent ponds and by skilful use of the levels in undulating lands, pond can be constructed without much expense of labour. A number of small, shallow ponds are much preferred to one or two large ones. From the economic point of view it is best to construct ponds which require little digging and a minimum artificial banking.

Special Efforts to Protect Biodiversity

The sub-Himalayan West Bengal and adjoining NE region is very rich in its biodiversity. Special efforts should be made for developing techniques for propagation and conservation of these species. It is noted with concern that an illegal trade exists in which natural collections are made from high altitudes and different habitats and these are exported as ornamental fishes to other countries where these can be exploited for development of new strains/varieties for commercial purpose and trade. Special efforts should be made to protect biodiversity. The importance and potential of Coldwater Ornamental Fishery in the sub-Himalayan West Bengal was well recognized and it was highlighted that since the State is promoting this sector in big way contributing to national and international trade in this field. It is strongly recommended that for indigenous varieties of ornamental fish only those species to be promoted for which breeding protocols are developed, wild collections to be discouraged to protect local biodiversity. It was recognized that trade in ornamental fishery in the State of West Bengal has significant economic benefits to society for which many species from hill regions are also exploited. The manpower support with the Department cannot police the river stretches for many irrational fishing practices like dynamiting in the rivers. Therefore, co-management approach is advocated, the local people may be trained in fishing methods, so that the dangerous fishing practices are discouraged.

Data Management

The improper database is one of the weakest links in formulation of management policies for fisheries and aquaculture sectors. Adequate and reliable database should be established, as it forms pre-requisite for plan of future course of action in this sector. Resource surveys conducted by scientific institutions, state fishery departments and local universities have indicated existence of a very large bio-diversity of species. The conservation of this biodiversity and its economical use is a priority area. Coordinated and special efforts are to be taken for conservation of some of the identified threatened indigenous fish species. We have to adopt advance stock assessment methodology by using appropriate software. Based on the generated data, appropriate management guideline will be provided in a way to ensure sustainable fishery stock. Local knowledge about fish abundance from the older generation has to be considered. Thus after generating a working data base on endangered fishes we can take up appropriate steps in technical and administrative lines to protect and preserve the endangered fish fauna.

Availability of Aquaculture Inputs

The availability of two most crucial inputs, quality seed and feed, both in time and space is the necessary pre-requisite for the success of any strategy for aquaculture development in every part of North Bengal.

Infrastructure Development

Infrastructure development both at production and post-harvest stage is a key requirement for accelerating the growth in aquaculture. However, it is strongly recommended that the facility to be further strengthened for creating hatchery, rearing and required support for mahseer fishery propagation.

Single Window Facility

To solve all aspects related to fish/shrimp aquaculture (including Govt. policy, technological assistance, credits, marketing information, availability of inputs etc.) may be brought under one governmental agency, which will ensure providing all sorts of service to the users/beneficiaries/farmers/entrepreneurs etc.

Leasing Policy

Specific policies should be framed to lease out different sized ponds/Jhora to fishermen co-operatives, group of fishers or individual fishermen, in order to have efficient management to exploit its full production potential.

Development of Eco-tourism

The Himalayan region is still 'Pristine' ecosystem and attracts national and international interest, therefore, development of eco-tourism should be promoted along with sustainable development of natural resources of the region including aquatic resources. Further, damage to our fragile Himalayan ecology should be avoided at any cost. The Ministry of Tourism should also allocate funds for any conservation effort. The construction of multipurpose dams and multipurpose reservoirs in the country and especially in Hill regions has received high priority at State and Central Governments. Their development should be with a multidisciplinary approach with emphasis on conservation and development of local resources including benefit to the local population. The experience during the last 50 years has shown that fisheries sector is a major component for generating production, income, livelihood, employment and quality of life. Therefore, it is recommended that integrated approach in conservation and development of fisheries, and eco-tourism should be adopted.

Development of Sport Fishing

Mahseer is a globally known as a sport fish of India. Internationally sport fisheries contribute significantly to the economy, sub-Himalayan West-Bengal and Sikkim have potential resources to develop sport fishery. But the scientific information suggest significant decline in Mahseer stocks in the local waters. It is felt that urgent steps be taken by the concerned authorities to check further decline. This calls for a concerted national effort of all institutions/organizations involved in riverine water use and fishery managers. The available state of art technologies should be utilized for setting up fish passes including ladders and mahseer hatcheries.

Holistic Development Approach

The upland aquatic resource management in West Bengal and Sikkim is different from other regions of the country. Their conservation, development and exploitation are different from tropical ecosystems of plains, thus special efforts are needed for holistic development of the hill streams/rivers and other associated ecosystems *viz.*, tals, lakes wetlands and small impoundments.

Budgetary Support

The fisher communities are generally poor and are in a disadvantageous position in the society. The majority of them belonged to the same area or to the vicinity of their enterprise making it easy for them to commute to their farm. The budgetary support is essential to meet the rising investment demands for research and development, infrastructure development, training and extension to cater the requirements of higher growth rate, emerging technologies and new possibilities in different aspects of aquaculture of North Bengal including the coldwater area. The upland regions have received marginal attention from the Planners so far, which has resulted slow development. The Government of India has already identified the northeastern states and hill regions as a priority area, similar support should be extended to sub–Himalayan West Bengal. Therefore, this area should be treated as priority area and higher fund allocation must be made for conservation, sustainable development to ensure better economy and improved quality of life for the people of the region. The scheme of "Jhora Fisheries" standardized by the Department of Fisheries has been successfully implemented in the upland regions of the state. However, it was recommended that science behind this scheme needs to be worked out on the basis of following items.

- ☆ Systems productivity, profitability, stability, sustainability and equity with due focus on gender issues
- ☆ Poor HDI as well as low per capita income across most of the constituent districts
- ☆ Long distance migration due to uncertainty in income seasonality

- ☆ Growing social disharmony and trans-border problems in the face of increasing livelihood challenges
- ☆ Mainstreaming linked women SHGs of North Bengal with required training and technological backstopping
- ☆ Gender disaggregated drudgery reduction
- ☆ Formulating appropriate mix of escape routes for provisioning of livelihood, protection of livelihood and promotion of livelihood to provide the target groups greater access to capital assets

Table 117: District-wise per Capita Income of North Bengal

District	*HDI*	*Per capita Income (in '000 Rs.)*
Darjeeling	0.65*	18.53
Jalpaiguri	0.53	16.74
Coochbihar	0.52	13.86
Uttar Dinajpur	0.51	11.18
Dakshin Dinajpur		14.58
Malda	0.44	14.78
West Bengal	0.61	16.02

This higher value over State is due to rapid growth of urban conglomeration of Siliguri city.

Strengthening of Extension Services

Considering the adoption gap of the aquaculture technologies in different parts of this sector, it is greatly emphasized that the extension machinery of the State Fisheries Departments are strengthened. It is recommended that researchable issues with regard to captive breeding, nutrition and feeds, of indigenous fishes be identified, which may be jointly addressed by the Zoology Faculty in North-Bengal University, Faculty of Fishery Science of West Bengal University of Animal and Fishery Sciences and State Department of Fisheries in West Bengal. The ICAR institutes *viz.*, CIFRI, Barrackpore and NRCCWF, Bhimtal to extend technical backstopping wherever required. It was recognized that manpower in hill fishery sector is limited. While stressing the need for organizing tailor made short-term training programmes it is strongly recommended that Zoology faculty in North-Bengal University, Faculty of Fishery Science of West Bengal University of Animal and Fishery Sciences should initiate some course work in Coldwater Fisheries within their existing programmes. It is also recommended that outside faculty may be involved in specialized topics.

R&D Support and Linkages

Strong linkages between research and development agencies (both in the State level as well as between National and State level) are pre-requisites for technology development and implementation as it is basically a multidisciplinary activity. GOI Research Funding Agencies should provide regularly some amount of fund for R& D work or to finance research project formulated by the State Fisheries Deptt. It is recommended that representatives of Fishers/Collectors, Trade, State Fisheries Department and Zoology Faculty in North-Bengal University should identify the indigenous species which can be traded on a sustainable basis, keeping in view the economic and ecological imperatives.

Human Resource Development

Human resource development programmes at different levels of researchers, development officials, extension officials, entrepreneurs and small and marginal farmers are vital to bring about the desired qualitative and quantitative improvement in aquaculture sector in the North Bengal. Distance education through print and electronic media and vocational courses and publication of brochure/pamphlets etc. in local languages may be given greater emphasis for spreading the technologies.

It is strongly recommended that under the Chairmanship of Hon'ble Minister of Fisheries, Aquaculture, Aquatic Resources and Fishing Harbours, Govt. of West Bengal, a working group may be constituted to oversee the development of fisheries sector of North Bengal including coldwater fisheries in the state and set priorities, frame policies, examine constraints and review implementation from time to time.

Stop Hydra Headed Approach

The challenge for both resource users and government is to carry out resource harvesting and full development of this ornamental fishery industry in a sustainable manner so that the natural resources are conserved and the risk of negative impacts and stock collapse is minimized. Both the goals for conservation and development can be achieved only through strategic planning and coordination among these various groups and stakeholders to avoid redundancy of developmental programmes and research projects. Thus, overall socio-economic development of pro-poor people may be formulated by taking all Government organizations, NGOs, SHGs and other stakeholders in a single forum. The current trouble is multiple individual agencies exists each trying to work on aquaculture and aquatic resource management without any concerted effort. So agencies like MPEDA, ICAR, Fishery Universities etc. work according to their own layout without consulting the State Fisheries Department which is intimately associated with all the stakeholders of Fishery sector. This hydra headed approach has resulted in confusion and retrogression of fisheries development in the state. It is a pity that even after 6 decades of independence an independent Fisheries Department in the Centre is yet to take shape. As the water resource *viz.* Rivers, Big water bodies are national treasures in terms of natural resource as they are not confined within the state boundaries, a piece meal approach towards the conservation would be of no use as because the pollution created upstream in some other states would effect the downstream states. Therefore, it is clear that the issue of fishery has a National appeal which can only be addressed effectively if tackled centrally. The achievements of some of the smaller South East Asian countries having GDP lower than that of ours have gone a long way due to an integrated approach on fisheries and water resources. Under such constraints the intended level of progress in terms of socioeconomic security for the stake holders, research activities are getting slowed down under multiple departmental activities and bureaucratic red tapes. Time has come to think seriously regarding a different line of approach giving the Department of Fisheries a much extended hand to deal with aquatic resource management as a whole and to size down the multiple decision making agencies into one *i.e.* Department of Fisheries.

Innovation in Design of Aquarium Accessories and Industry Set Up are the Need of the Day – But............

State industrial policy document classifies Agro and food processing sector as one of the thrust areas. However on the contrary the same policy document classifies aquaculture projects under the "Negative list of industries".

Incentives and subsidies enjoyed by the large scale (above 1 crore) as well as small scale industries (below 1 crore) under West Bengal incentive scheme 2004 as follows:

- ☆ State capital investment subsidy.
- ☆ Interest subsidy.
- ☆ Waive of electricity duty.
- ☆ Additional incentive on employment generation.
- ☆ Refund of stamp duty and registration fee.
- ☆ Subsidy for conversion for use of piped gas.
- ☆ Subsidy for quality improvement in the small-scale sector.

Ornamental fishery has a very bright prospect not only in the domestic market but has tremendous export potential as well. This can provide the economic independence to the women folks with out much investment and running around. So this sector has to be encouraged by providing marketing facility. This present lack of marketing system is hampering the development of a very promising sector both in large scale and small scale of women groups. The allied sector of ornamental accessories is also a potential sector for development. Appropriate measures should be encouraged for artificial propagation of such identified species so that economic benefits can be derived without affecting their natural abundance. Marine Products Export Development Authority (MPEDA) should provide adequate facility for marketing which is their primary field of activity in association with the DOF, GOWB. Since the nearest international airport to the collection sites is located at Kolkata, the trades of this species have thrived in West Bengal. Few traders and exporters of Kolkata regulate this market. This industry being unorganized and yet having considerable export potentials would require checks and balances to confirm international standard thereby maintaining purity of character. At present the export from this sector is based on the natural collection of indigenous species. This may result in depletion of natural stock besides creating faunistic imbalance. The domestic market on the other hand is with non-descriptive breed as a result of indiscriminate breeding operation as well as import of non-certified seed as well as brood stock. Besides, the present operations are not driven by technology. The non-scientific character of the operations has its root in the poor farm gate pricing of the produce, which is controlled by few traders, thereby not permitting adequate investment commensurate with scientific farming. Uncontrolled breeding/inbreeding in the sector has resulted in the dilution of the characteristic of the species. This, coupled with inadequate consumer awareness has led to an invasion of scrap ornamental fishes in the secondary market.

Bibliography

Anon.1992. Participatory Rural Appraisal (PRA) for Agricultural Research at Aruppukottai, Tamil Nadu. Tamil Nadu Agricultural University and International Institute for Environment and development, London.

APHA, 1995. Standard methods for the examination of water and waste water Analysis, 18th Ed. *Amer. Pub. Hlth. Ass., Awwa, WBPC. Washington DC.*

Barat, S., Prithawiraj Jha and R.F. Lepcha. 2005. Bionomics And Cultural prospects of Katli, *Neolissocheilus hexagonolepis* (McClelland) in Darjeeling district of West Bengal. In Coldwater Fisheries Research and Development in North –east region of India (Eds. B. Tyagi, Shyam Sunder and M. Mohan). NRCCWF, Bhimtal. Vikrant Computers Haldwani: 66-69.

Chacko, P.I., G.K. Kuriyan and S. Thyagarajan. 1954. A survey of the fisheries of the Cauvery river. Contrib. Feshwater Fish. Biol. Station, Madras 12:1-19.

Chakraborty C., 2007a. Fisheries of Wetlands: Issues of resurrection. In the Souvenir of 26th Bi-Annual meet-2007 organized by State Fisheries Officers' Association, West Bengal during 19-21st July, 2007 at Kolkata.

Chakraborty C., 2007b. Coldwater Fisheries of West Bengal: The Present Scenario. In the Souvenir of 26th Bi-Annual meet-2007 organized by State Fisheries Officers' Association, West Bengal during 19-21st July, 2007 at Kolkata.

Chowdhury, H., Singh,S. B. and Sukumaran, K. K. 1966. Experiments on large scale production of fish seed of Chinese carps, *Ctenopharingodon idella* (C&V) and the silver carp, *Hypopthalmicthys molitrix* (C&V). By induced breeding in pond in India. *Prooc. India Acad. Sci.,* 53 :80 – 95.

Council of Scientific and Industrial Research, 1962. *The Wealth of India, Raw Materials Vol. IV, supplement Fish and Fisheries,* New Delhi.

Das, P. and Pandey, A. K. 1998. Current Status of fish germplasm resources of India and stategies of conservation of endangered species, pp. 252 – 272. *In*: Ponniah, A. G., Das, P. and S. R. Verma (eds), *Fish Genetics and Biodiversity Conservation*. Nature Conservators, Muzaffarnagar.

Day, F. 1967. *The fishes of India. Vol. I.* Jagmander book agency, New Delhi.

Dehadrai, P. V., Das, P. and Verma, S. R. 1994. Threatened fishes of India. *Nature Conservators.* Muzaffarnagar.

Dubey, G.P., 1978. Survey of the waters of Bhutan, Physiography and Fisheries potential. FAO, Rome. 38p.

Ghosal, D.K. 2002. Scope for development of coldwater fisheries in the state of West Bengal. In *Highland Fisheries and Aquatic resource Management* (Eds. K.K. Vass and H.S. Raina). NRCCWF, Bhimtal. Yugantar Prakashan Pvt. Ltd., New Delhi: 73-75.

Hora, S. L. 1952. The Himalayan Fishes, Himalayan, 1(i): 66-74

Jhingran, V. G. 1982. Procurement of stocking material. pp. 355-382. In: *Fish and Fisheries of India,* Hindusthan publication Corporation (India). Delhi.

Jhingran, V.G and K.L. Sehgal. 1978. *Coldwater fisheries of India.* Inland Fisheries Society of India, Barrackpore: 1-239.

Joshi, C.B. and Raina, H. S. 1997. Impact of habitat changes on Mahseer fishery in India upland with special references to Kumaon Himalayas. pp. 88 *In*: Vass, K. K. and M. Sinha (eds.). *Changing Perspective of Inland fisheries,* Inland Fisheries Society of India, Barrackpore.

Karmakar, A., Bandyopadhyay, A. and Mukherjee, M. 2000. Freshwater prawn, *Macrobrachium rosenbergii* post larvae production by modified management protocols. pp 157 – 161 *In*: B. B. Jana, R. D. Banerjee, B. Guterstam and J. Heeb (eds.). *Waste Recycling and Resource Management in the developing world.*

Kirchhofer, A. and Hefti, D. 1996. *Conservation of endangered freshwater fish in Europe,* Birkhauser Verlag, Basel, Switzerland.

Mahanta, P.C. and S. Sunder, 2006. Coldwater fisheries in Himalayan uplands with special emphasis on North Bengal. Compendium on An approach to the State level seminar in prospects and potentialities on Coldwater fisheries in West Bengal; Held on March 21-22, 2006 at Siliguri, Darjeeling. P13-23.

Maitland, P. S. 1993. Conservation of freshwater fish in India. pp. 349 – 364 *In*: B. R. Singh (ed.) *Advances in Fish Research.* Narendra Pub. House, New Delhi.

Master, L. M. 1990. The imperiled status of North America aquatic animal. *Bioderversity Network News,* 3: 7 – 8.

Meffe, G. 1987. Conserving fish genomes: philosophies and practices. *Environ. Biol. Fishes,* 18: 3 – 9.

Menon, A. G. K. 1996. Endangered, vulnerable and rare freshwater fishes in India and their conservation. pp 26-28. *In*: P. Das., A. G. Ponniah., K. K. Lal and A. K.

Pandey (eds.) *Symposium on Fish genetic and Biodiversity conservation for sustainable production.*

Menon, A.G.K., 1999. *Checklist of Freshwater Fishes of India*, Zoological Survey of India.

Miller, R.R., Williams, J. D. and. Williams, J. E. 1989. Extinction of North American fishes during past century. *Fisheries (Bethesda)*, 14(6): 22 –38.

Mukherjee, Madhumita and G. Sarkar. 2005. Endangered Fishes of West Bengal, with a special reference to North Bengal on research, restoration, and Future Plan of action. Published by Deptt. of Fisheries, Aquaculture, Aquatic Resources and Fishing Harbour, April 2005.

Mukherjee, Madhumita. 2002. Problems and prospects of coldwater fisheries in West Bengal. In *Highland Fisheries and Aquatic Resource Management* (Eds. K. K. Vass and H.S. Raina). NRCCWF, Bhimtal. Yugantar Prakashan Pvt. Ltd., New Delhi: 86-90.

Ng, H.H. 2003., A revision of the south Asian Sisorid catfish genus Sisor (Teleostei: siluriformes), *Journal of Natural History*, 2003, 1-13.

Pavlov, D.S. 1993. Strategies for preserving rare and endangered fishes. *J. Ichthyol.*, 33(1): 109 – 127.

Petr.T., 1999. *Coldwater Fish and Fisheries in Bhutan*. FAO Fisheries Technical Paper. No. 385: 6-12. FAO, Rome.

Ponniah, A. G., Pandey, A.1999. Endangered Fish species: Measures for Rehabilitation and Conservation. *Fishing Chimes* 19(6): 31 – 34.

R.F.Lepccha, A. Mandal and S. Pradhan. 2006. Studies on coldwater fisheries in hill area of Darjeeling District. Compendium on An approach to the State level seminar in prospects and potentialities on Coldwater fisheries in West Bengal; Held on March 21-22, 2006 at Siliguri, Darjeeling.p 77-80.

Rao, Andrew Arunava, 2006. Indigenous Coldwater Fish potential for ornamental Fish exports. Compendium on An approach to the State level seminar in prospects and potentialities on Coldwater fisheries in West Bengal; Held on March 21-22, 2006 at Siliguri, Darjeeling.p 50-54.

Raychaudhuri Bikash. 2007. *The Rabhas of North Bengal*. pp 85.

Schonhut, M. and U. Kievelitz, 1994. Participatory Learning Approaches– Rapid Rural Appraisal, Participatory Appraisal, an introductory guide. TZ-Verl-Ges., pp 183.

Sen, P.K. and Biswas, S.N., 2006. *Developmental strategies of Coldwater fisheries in Darjeeling Hill areas of West Bengal-An Overview*. Compendium on An approach to the State level seminar in prospects and potentialities on Coldwater fisheries in West Bengal; Held on March 21-22, 2006 at Siliguri, Darjeeling. P24-49.

Shresha, J., 1999. Coldwater fish and fisheries in Nepal. In: *Fish and Fisheries at Higher Altitudes*: Asia (T. Petr, ed.) : 13-40. FAO Fisheries Tech. Paper. No. 385. Rome, FAO.

Shrestha, T.K. (1990) Rare Fishes of Himalayan Waters of Nepal, Journal of Fish Biology (1990) 37 (supplement a). Academic Press, London, pp.213-216.

Shrestha, T.K. 1997. The Mahseer in the Rivers of Nepal disrupted by dams and Ramching Strategies. R. K. Priaters, Kathmandu, Nepal.

Smith, A. H. and Chesser, R. K. 1981. Rationale for conserving genetic variation of fish gene pools. *Fish gene pool Ecol. Bull* (Stockholm), 39: 13 – 90.

Talwar, P.K. and A.G. Jhingram, 1991. Inland Fishes. Vol 1 and 2. Oxford and IBH Publishing co., Ltd. New Delhi. 1158p.

Tekriwal, K.L. and Rao, A.A., 1999, Ornamental aquarium Fish of India (Waterlooville: Kingdom Books), 144 pp.

Townsley, P. 1993. Rapid Appraisal Methods for Coastal Communities. BOBP, Madras, pp 1.

Index